MINING IN THE ARCTIC

PROCEEDINGS OF THE SIXTH INTERNATIONAL SYMPOSIUM ON MINING IN THE ARCTIC
NUUK, GREENLAND, 28-31 MAY 2001

Mining in the Arctic

Edited by

Hans Kristian Olsen, Lida Lorentzen & Ole Rendal
Bureau of Minerals & Petroleum, Nuuk, Greenland

CRC Press
Taylor & Francis Group
Boca Raton London New York

CRC Press is an imprint of the
Taylor & Francis Group, an **informa** business
A BALKEMA BOOK

The texts of the various papers in this volume were set individually by typists under the supervision of either each of the authors concerned or the editor.

First published 2001 by Swets & Zeitlinger

Published 2022 by:
CRC Press
P.O. Box 447, 2300 AK Leiden, The Netherlands
e-mail: Pub.NL@taylorandfrancis.com
www.crcpress.com – www.taylorandfrancis.com

ISBN-13: 978-90-5809-177-2 (hbk)

Mining in the Arctic, Olsen, Lorentzen & Rendal (eds), © 2001 Swets & Zeitlinger, Lisse, ISBN 90 5809 177 5

Table of contents

Mining engineering and mine design in permafrost regions

Mining in the Arctic, Olsen, Lorentzen & Rendal (eds), © 2001 Swets & Zeitlinger, Lisse, ISBN 90 5809 177 5

FOREWORD

The 6th International Symposium on Mining in the Arctic - as the previous ones held in Fairbanks, Alaska in 1989 and 1992; St. Petersburg, Russia in 1994; Longyearbyen, Svalbard in 1996; Yellowknife, Northwest Territories, Canada in 1998 - is aimed at:

- Advancing the general knowledge of issues relating to mining in the Arctic
- Improving the exchange of information between the Arctic regions
- Encouraging research co-operation and communication between the mining establishments of the countries concerned
- Increasing the understanding of the impact of mining on an extremely sensitive environment

The papers presented at the Symposium cover a wide variety of topics related to:

- Mineral exploration and exploitation
- Mining engineering and mine design in permafrost regions
- Environmental impact of mining in the Arctic
- Government administration and legislation related to mining

The weighting of the themes that are being addressed at the 6th Symposium reflect the steadily increasing realisation that exploration and mining of Earth's mineral resources must be conducted in an environmentally responsible manner and be of long-term benefit to the communities involved.

A workshop on "Control Strategies for Acid Mine Drainage (AMD) in Arctic Regions" also emphasises the importance of protecting the sensitive environment of the Arctic. The workshop aims to be a forum for presenting and discussing future AMD control strategies in Arctic regions based on present knowledge and the following basic approaches:

- Control of acid generation reactions
- Control of migration of contaminants
- Collection and treatment

The papers related to mining engineering, mineral exploration and other issues emphasize the challenges to exploration and mining posed by the special climatic conditions, infrastructure and geomorphic characteristics of the Arctic.

As organisers of the 6th International Symposium on Mining in the Arctic, we wish to thank those organisations and persons who have offered substantial support throughout the planning:

- Deputy Minister Hans Kristian Schønwandt and staff of the Bureau of Minerals and Petroleum, Government of Greenland, for administrative support
- Greenland Resources A/S, for major financial support
- Professor Sukumar Bandopadhyay, Head of the Department of Mining and Geological Engineering at the University of Alaska, for invaluable help and advice
- Greenland Incoming for logistical support

Finally, we wish to express our gratitude for generous financial contributions from the following sponsors:

- Crew Development Corporation, Canada
- Falconbridge Ltd., Canada
- Greenland Resources A/S, Greenland
- Gryphon Metals Corporation, Canada
- New Millennium Resources N.L., Australia
- Nalunaq I/S, Greenland
- Nunafonden, Greenland
- Nunaminerals A/S, Greenland
- Platinova A/S, Canada
- Raw Materials Group, Sweden
- Store Norske Spitsbergen Kulkompani AS, Svalbard

SESSION CHAIR

Sukumar Bandopadhyay
University of Alaska, Fairbanks, USA

Bob Jacko
Cominco Ltd., Kimberley, Canada

Hans Kristian Olsen
Bureau of Minerals & Petroleum, Greenland

John E. Udd
Mining Laboratories, CANMET, Canada

Johannes Vik
Commissioner of Mines at Svalbard, Norway

INTERNATIONAL ARCTIC MINING ADVISORY COUNCIL

CANADA

John E. Udd
Chairperson, 1996-1998
Mining Research Laboratory
Canmet, Ottawa

A. J. (Tony) Keen
Cominco Ltd. Vancouver

B. Ladanyi
École Polytechnique, Montréal

Raj Singhal
Revenue Canada, Calgary

FINLAND

Raimo Matikainen
Helsinki University of
Technology, Helsinki

Anti Mikkonen
Siilinjärvi Works and Mine

GREENLAND

Hans Kristian Olsen
Chairperson, 1998-2001
Bureau of Minerals and
Petroleum, Nuuk

NORWAY

Arne M. Myrvang
Chairperson, 1994-1996
Norwegian Institute of
Technology, Trondheim

John Utsi
Store Norske Spitsbergen
Kulkompani AS,
Longyearbyen, Svalbard

Johannes Vik
Commissioner of Mines,
Svalbard

PEOPLE´S REPUBLIC OF CHINA

Ma Guang
Hengyang Institute of Technology, Hunan

Wang Changhan
Hengyang Engineering College, Hunan

RUSSIA

Victor L. Yakolev
Mining Institute of the North, Yaknisk

Clement N. Trubetskoy
Ipcon, Moscow

V.Y. Izaxon
Mining Institute of the North, Yakutsk

Nicholai N. Melnikov
Kola Mining Institute, Apatity

Vasili J. Polijuk
Alamazzolotoprogress, Moscow

Yuri I. Anstratov
Institute for Geology and Prospecting, Moscow

V.S. Litvinenko,
St. Petersburg Mining Institute, St. Petersburg

UNITED STATES OF AMERICA

Sukumar Bandopadhyay,
Chairperson, 1986-1992
University of Alaska, Fairbanks.

John F. Sims
Usibelli Coal Mine, Fairbanks

Earl H. Beistline
Mining Consultant, Fairbanks

L. Michael Kaas
U.S. Department of Interior-BLM, Washington

Michael G. Nelson
EIMCO, Salt Lake City

Joseph B. Seymour
NIOSH, Spokane

Environmental impact of mining in the Arctic

Mining in the Arctic, Olsen, Lorentzen & Rendal (eds), © 2001 Swets & Zeitlinger, Lisse, ISBN 90 5809 177 5

Environmental regulation of mineral exploration in Greenland

P. Aastrup & C. M. Glahder
National Environmental Research Institute, Department of Arctic Environment, Denmark

ABSTRACT: Mineral exploration in Greenland is regulated by the Bureau of Minerals and Petroleum in Nuuk with NERI acting as an environmental consultant. The regulation is based on a general set of rules for driving, flying, storage of fuel etc. Examples of environmental effects of human activities are given for caribou, geese and vegetation.

In addition to the general rules a number of areas have been designated to be of special importance for wildlife. In sensitive time periods for wildlife, mineral activities require specific approval from the authorities in these areas. All regulations can be found on the Internet where it is also possible to zoom in on any area to find out which kind of regulations are in force.

1 INTRODUCTION

Mineral exploration in Greenland is regulated by the Bureau of Minerals and Petroleum (BMP), the Greenland Home Rule Government. The National Environmental Research Institute (NERI) conducts research and advises BMP on issues related to environmental management of mineral exploration and exploitation. The regulation of mineral activities is based on general rules and rules for diamond drilling, driving, explosives, and access to important areas for wildlife and preserved areas (BMP 1999).

The general environmental rules on mineral exploration state in broader terms that e.g., wildlife shall not be disturbed unnecessarily, vegetation shall not be damaged unnecessarily, pollution shall be reduced as much as possible, combustible waste shall be burnt or transported to town dump sites, whereas hazardous waste shall be delivered to a municipal receiving system.

More specific environmental rules are given for areas important to wildlife. These areas cover seabird colonies, moulting areas for eiders and geese, calving grounds for caribou and muskoxen, walrus haul-out site, etc. For each type of area the sensitive periods are given. During these periods all mineral activities have to be approved by BMP. The areas are shown on maps which can be found also on the BMP homepage (www.bmp.gl).

In this paper we describe in more detail the designated areas important to wildlife and how the material can be used by mining companies and administrators. We give examples on the designation of areas, determination of sensitive periods and disturbance effects on caribou and geese. Also, examples are given on effects on vegetation due to e.g., driving.

It is our hope that this paper will make it easier to understand the background for both general environmental regulations and for specific approval of activities in areas designated as important to wildlife.

2 DISTURBANCE REACTIONS OF CARIBOU

Caribou (*Rangifer tarandus*) are found in West Greenland and now the total population is estimated at 30.000-40.000 individuals. Pregnant Caribou cows aggregate on calving grounds that are used year after year. Calving grounds are often early snow free areas that offer nutritious forage, which is highly important after the winter.

Calving take place in June. Studies of caribou behaviour have shown (Aastrup, 2000) that the awareness and sensitivity to disturbance is highest during calving (Fig. 1). It was also shown that groups with calves were more aware than groups without calves (Fig. 2).

Calving areas are often protected to allow the cows and their calves to reconstitute after the winter and the birth. The pregnant females have often been deprived of resources after the winter and it is important, that calves as soon as possible can gain weight in order to be able to withstand climatic and other stress factors.

Disturbances influence the energy balance negatively by increasing energy demanding activities and decreasing food intake. This means that Caribou that have experienced heavy disturbances during summer, may not have been able to accumulate fat to survive the winter. Similarly, a high level of disturbance during winter will lead to unnecessary depletion of energy and fat.

It is generally agreed (e.g. Anon 1993) that calving areas deserve special protection during calving time.

Some sex- and age groups of Caribou are able to habituate to some disturbances. It is often seen that bulls forage close to roads and buildings. It is not seen, however, that pregnant females and new born calves stay in such areas. Experiences from Alaska and Norway show that caribou tend to avoid areas along roads power lines and tourist resorts (Cameron *et al.* 1992; Vistnes & Nellemann 2000).

The examples from Greenland confirm that caribou are more sensitive to disturbances during calving time and that groups with calves are more sensitive to disturbances than groups without calves.

The study has provided: 1) documentation that Greenland caribou almost always respond to disturbance and behave much like caribou in e.g. Norway and Svalbard, 2) support for the assumption that caribou in Greenland are more sensitive during calving than in summer and to some degree also after hunting, and 3) documentation that the effect of disturbance is stronger during calving than in summer and in autumn after hunting.

The general rules for field work include precautions to avoid unnecessary disturbance of caribou in Greenland by stipulating that: Flights with fixed-winged aircrafts shall be above 500 m above terrain and that activities within calving ranges during the period May 20th – June 20th shall be approved by the Bureau of Minerals and Petroleum.

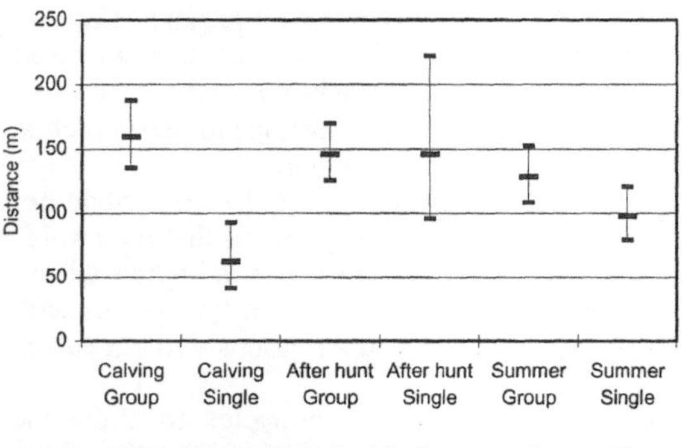

Figure 1. Flight threshold distance for caribou at different seasons in Greenland.

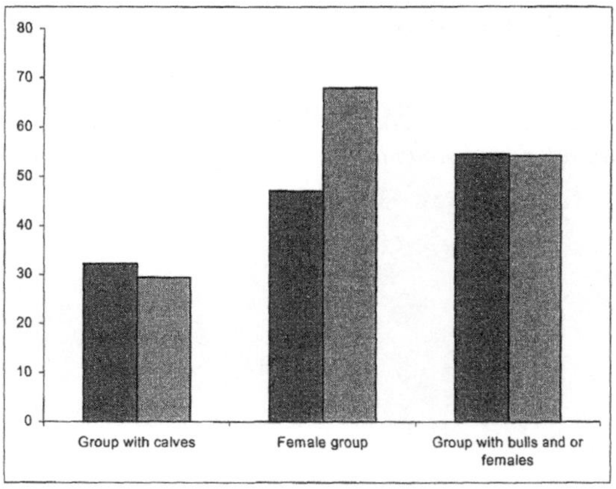

Figure 2. Frequency (%) of curiosity behaviour in groups with calves, female groups without calves and groups with both females and bulls. Dark grey columns Kangerlussuaq, light grey columns Nuuk (n=168)

3 DISTURBANCE REACTIONS OF GEESE

3.1 *Goose biology*

A total of six different goose species are breeding in Greenland (Boertmann & Glahder 1999). Of these, sensitive periods and disturbance effects have been studied primarily in Greenland White-fronted *Anser albifrons flavirostris*, Pink-footed *A. brachyrhynchus*, and Barnacle goose *Branta leucopsis*.

All goose species visit Greenland only during the summer months. Winters are spent in either West Europe or North America. The geese arrive in West Greenland during May and the first three weeks the geese feed extensively to fill up their fat depots. These depots are depleated during their spring migrations which cover thousands of km's. It is of outstanding importance to the breeding geese to fill up their depots so that they are able to produce and incubate their eggs, and defend their goslings. Immediately after the four weeks of incubation, during which period the pair hardly feed, the female goose might have lost about one third of her body weight (Ankney & MacInnes 1978).

The geese moult for about four weeks in July. During this period they are not able to fly because they have shed all remigies. The geese are able to maintain their body weight during moult, even though up to one third of the body protein is shed with the feathers (Hohman et al. 1992). This is, in part, due to the geese saving energy because they are not able to fly. Relative to the base energi consumption, flying demands 12.0 times more energy, while running takes 2.8 times more (Ravelin & LeFebvre 1967, Miller et al. 1994, Fig. 3). During August and September the geese build up their fat depots for the autumn migration back to the wintering grounds.

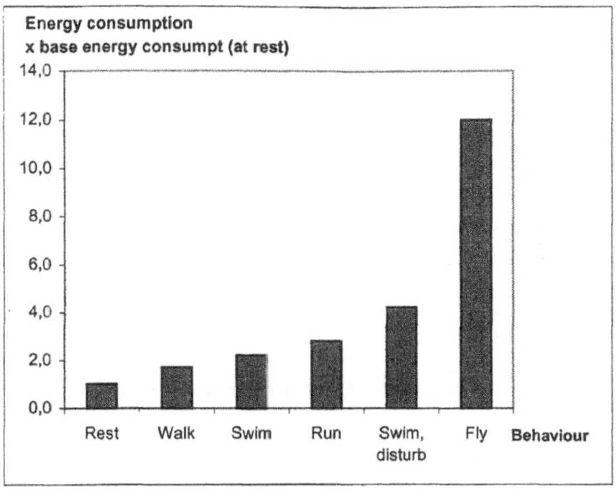

Figure 3. Energy consumption at different goose behaviours. When geese are disturbed they use from 3 (run) to 12 (fly) times more energy than at rest.

Reactions to disturbances are parts of the every-day life of the geese. They are therefore adapted to disturbances from natural factors, e.g., hunting from Arctic foxes *Alopex lagopus* and White-tailed eagles *Haliaeetus albicilla*. Disturbances from human activities such as mineral explorations come on top of the natural disturbances. If these disturbances are extensive they can affect the geese's feeding abilities and defence of nests and goslings. Major activities can impact on the whole population if they disturb important areas such as major spring staging areas, large breeding colonies, and major moulting grounds.

3.2 *Spring staging areas*

Spring staging areas have been studied in the Greenland White-fronted Goose population breeding in West Greenland between 64°N and 72°N. Ten major staging areas, holding 80-85% of the geese counted at aerial surveys, are situated in the early ice-free area between 66°30'N and 69°N (Glahder 1999a, Glahder et al., in prep.). The geese stage at these areas for an average minimum period of 11.2 days in May (Glahder et al. 1999a, in prep.). During staging the geese feed for nearly 80% of a diurnal period (Glahder et al. 1999b). Extensive disturbances at these major staging areas during the staging period will decrease the feeding period of the geese, which in turn can impact on the number of pairs that are able to breed successfully.

During the staging period the geese will stay alert when walking persons are closer than 700 m, and they will fly when persons are 200-500 m away. Fixed-wing aircraft will flush the geese at distances of 700-1000 m above ground (Glahder et al. 1999b).

3.3 *Breeding grounds*

Most goose species breed in single pairs with one km or more between nests. Only Barnacle Goose breed in colonies with nests situated on cliffs. About 100 such colonies are distributed along the East Greenland coast (70°-79°N) (Boertmann & Glahder 1999, Ogilvie et al. 1999). So, in most cases mineral exploration will affect only few nesting pairs during the breeding season. This is fortunate because disturbances during incubation would weaken the female goose if flushed too often; also, the eggs might be too cold and be exposed for too long to predatores like Arctic foxes, Ravens *Corvus corax* and Glaucous gulls *Larus hyperboreus*.

3.4 *Moulting areas*

Moulting geese, being temporarily flightless, are relatively tied to their moulting grounds for about a month in July. Therefore, a moulting ground must fullfill the geese's need for sufficient feeding areas and for refuge areas, such as lakes or streams. Disturbance reactions of Greenland White-fronted Geese have during 1999 and 2000 been studied (Glahder & Walsh, unpubl.) at Naternaq (68°20'N, 52°W), the most important moulting ground in West Greenland (Glahder 1999b).

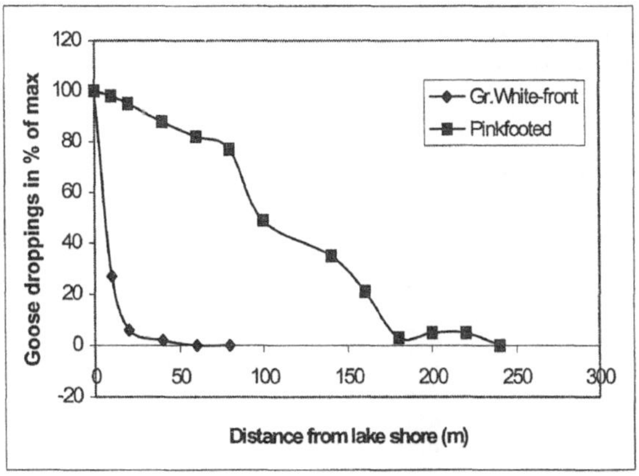

Figure 4. Two goose species' utilization of their moulting ground expressed as goose droppings pr. m^2 in % of maximum numbers. Pink-footed Goose (Madsen & Mortensen 1987), Greenland White-fronted Goose (Glahder & Walsh, unpubl.).

Because moulting geese are only able to escape predators by running to the refuge waters the geese tend to stay rather close these. Figure 4 shows distances to refuge lakes in two different goose species measured from number of goose droppings per m^2.

The two species show marked behavioural differences, but in generel most geese will stay not further than 100 m's from the lake shore. Taking in consideration that the average speed of a running goose over short distances is 5.0 km/h (SD=1.3, n=4), it is not surprising they stay close to the refuges.

In the study, a person walked directly towards a flock of geese and both parties were able to see each other. The geese became alert at an average distance of 782 m (SD=235, n=6). When the person walked on the geese ran away at an average distance of 482 m (SD=160, n=14) (Fig. 5).

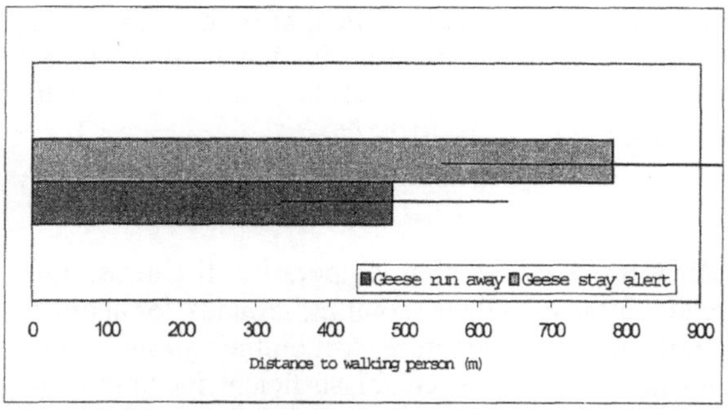

Figure 5. Behavioural reactions of moulting Greenland White-fronted Geese.

In most cases the geese left their moulting area and ran several kilometers over land. The majority of flocks had returned to their moulting grounds three days after the scare. Yet, one third of the flocks did not return until five day or more after the scare (Fig. 6).

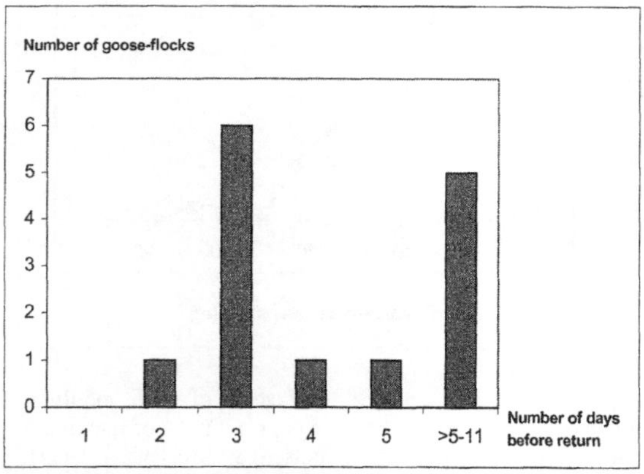

Figure 6. Number of days before a flock of moulting Greenland White-fronted Geese return to their moulting ground following a disturbance on day zero.

The reaction of moulting geese to helicopters was studied in Jameson Land, East Greenland (Mosbech & Glahder 1991). This area holds about 5000 moulting Pink-footed Geese and 5000 Barnacle Geese. The geese became alert to approaching helicopters at distances between 2.5 and 9 km. The distances were dependant on both goose species and helicopter type (Table 1).

Table 1. Distances (in km) to approaching helicopters where moulting geese showed alert reactions. Two goose species and two helicopter types were studied in Jameson Land, East Greenland.

Helicopter type	Pinkfooted Goose	Barnacle goose
Small-Bell 206	4.4 (1.0-6.5)	2.6 (0-5)
Large-Bell 212	8,7 (3-23)	9.1 (7.5-15)

Variations in distance (km) shown in brackets.

When the helicopters came closer to the geese they ran away as described for Greenland White-fronted Geese being disturbed by a walking person. It was demonstrated that helicopter disturbances affected goose behaviour, so that the time devoted for feeding decreased when disturbances increased (Fig. 7).

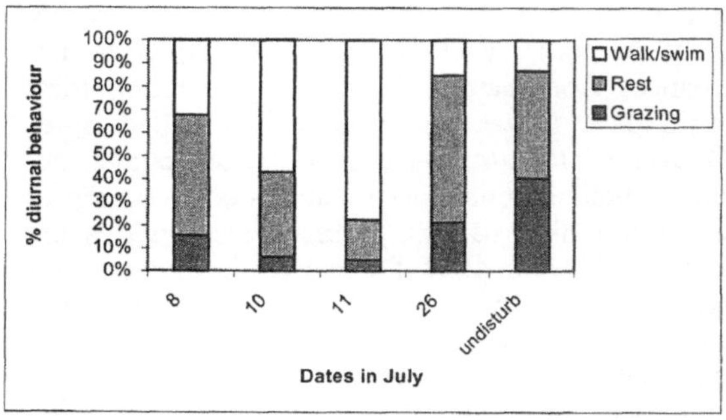

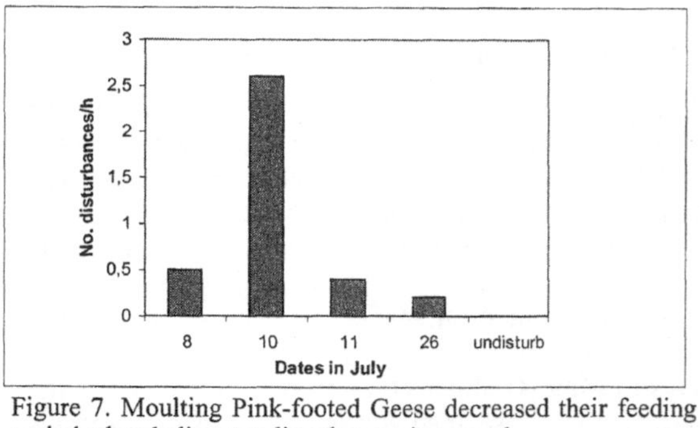

Figure 7. Moulting Pink-footed Geese decreased their feeding period when helicopter disturbances increased.

4 DAMAGE TO VEGETATION

Arctic vegetation exists in a cold climate with a short growth season and often there is also limited access to water and nutrients. Regeneration time is very long, which means that tracks from driving or other activities may stand for decades or centuries.

Damage to vegetation can be caused either directly by stripping off vegetation cover or by altering the physical characteristics of the soil.

Fig. 8 shows changes in the vegetation cover in Greenland dwarf shrub heaths in caused by driving related to seismic studies in Jameson Land, northeast Greenland.

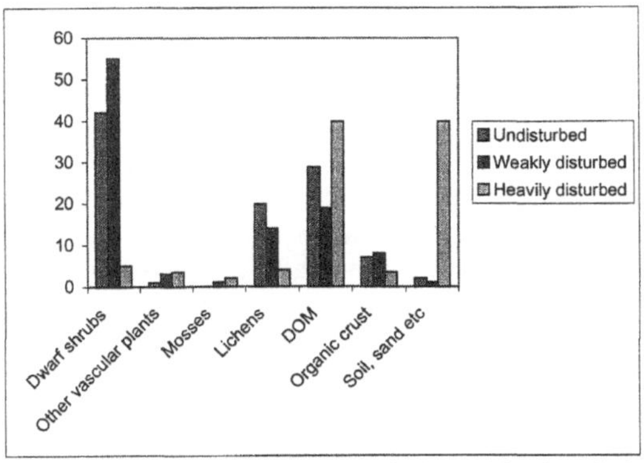

Figure 8. Characteristics of undisturbed, weakly disturbed and heavily disturbed dwarf shrub heath 6 years after disturbance. Efter Strandberg (1997).

The figure illustrates, that disturbed vegetation remains very different from the original vegetation even after a long time of recovery. This is seen from the decrease in the amount of dwarf shrubs and lichens, in the increase in the cover of vascular plants which initiate a new succession of plant growth, and the amount of dead organic matter (DOM) and soil , sand etc. Fig. 9 shows a track in *Cassiope tetragona* heath in Jameson Land, Northeast Greenland.

Effects of driving and other destructive activities can be minimized by securing that: 1) soil is well drained, 2) soil is frozen to minimum depth of 20 cm, 3) thickness of snow cover is at least 20 cm and, finally 4) by reconnaissance of proposed driving routes taking care that the most sensitive areas are protected.

Figure 9. Tracks from driving related to seismic activities in Jameson Land, Northeast Greenland.

Vegetation is an important element in the landscape, and in addition to this, vegetation is important forage for wildlife and it reduces the risk of soil erosion.

5 AREAS IMPORTANT TO WILDLIFE

Greenland is a very large country and important ranges for wildlife are found scattered all over the ice free part. Caribou and muskox (*Ovibos moschatus*) are important terrestrial mammals and important birds are geese and seabirds. Generally the vegetation layer is thin and most well developed in the southern and western parts. Above we have presented some of the scientific studies which have been the base for the delimitation of areas and periods important to Caribou and geese in Greenland. Areas important to other animal species have been delimitated in the whole of Greenland as well, and the different species appear from the legend in Figure 10.

Information about important areas and their sensitive periods is compiled in a GIS database, part of which is accessible through the Internet. Here, the areas are shown on 16 large-scale maps from which it is possible to zoom-in for more detailled information. An example of one of the detailled maps is shown in Figure 10. Here Caribou calving ranges are shown together with two different license areas. Mining companies that e.g., are preparing for their field season, can enter the BMP homepage, www.bmp.gl/E/EB2_minerals/EB2_40aa_20nn_fiel dwkreport-wildlifeintr.html, to get quick information about areas where specific rules may apply. The themes of the maps are:

1) Preserved areas
2) Ramsar areas
3) Towns and settlements
4) Areas with special restrictions
5) Sea bird colonies
6) Muskox areas
7) Caribou areas including calving grounds
8) Breeding, moulting and staging areas for geese
9) Themes from maps in "Rules for field work and reporting regarding mineral resources (excluding hydrocarbons) in Greenland".

Each signature on the maps refers to a specific species, the species' behaviour (breeding, calving, moulting, denning, etc.), and the sensitive period. During the sensitive periods all mineral activities inside important areas have to be approved. However, fixed-wing aircrafts flying a minimum of 500 m above ground or sea level does not require approval. Because reactions to disturbances differs among species and their annual cycle, restrictions in mineral activities differs. For example, around colonies of breeding Brünnich's guillemots there is a safety zone of 5 km, while this is only 200 m around other seabird colonies. Helicopters are allowed to pass some areas if they hold a minimum of 500 m above ground or sea level, whereas such flights across other areas are restricted to only one crosssing every fourth day.

6 FINAL REMARKS

The aim of the "Rules for field work and reporting regarding mineral resources (excluding hydrocarbons) in Greenland" is to protect wildlife and vegetation so that the effects of mineral activities are held at an acceptable level. Above we have presented some of the scientific studies which have been the base for the designation of areas and periods important to Caribou and geese in Greenland. Areas important to other animal species have been designated in the whole of Greenland as well, and the different species appear from the legend in figure 10.

The philosophy behind the rules is that almost anything is possible but that care has to be taken by following the general rules. A number of areas important to wildlife have been designated, and in most of these areas any activity requires approval by the BMP in Nuuk.

their licence areas, and when the sensitive periods applies.

The easiest way for all parties is, if the mining company can plan their activities so that exploration is avoided in the important wildlife areas during the sensitive periods. If this is not possible the company must apply for approval.

The basis for the designation of areas important to wildlife is studies perfomed by NERI together with studies on relevant litterature. Aerial surveys over huge areas of Greenland have delimited the important areas, while studies on the ground have revealed the sensitive periods and the sensitive reactions of the animals. The latter knowledge is of importance both when the areas are delimitated and when specific mineral activities are approved.

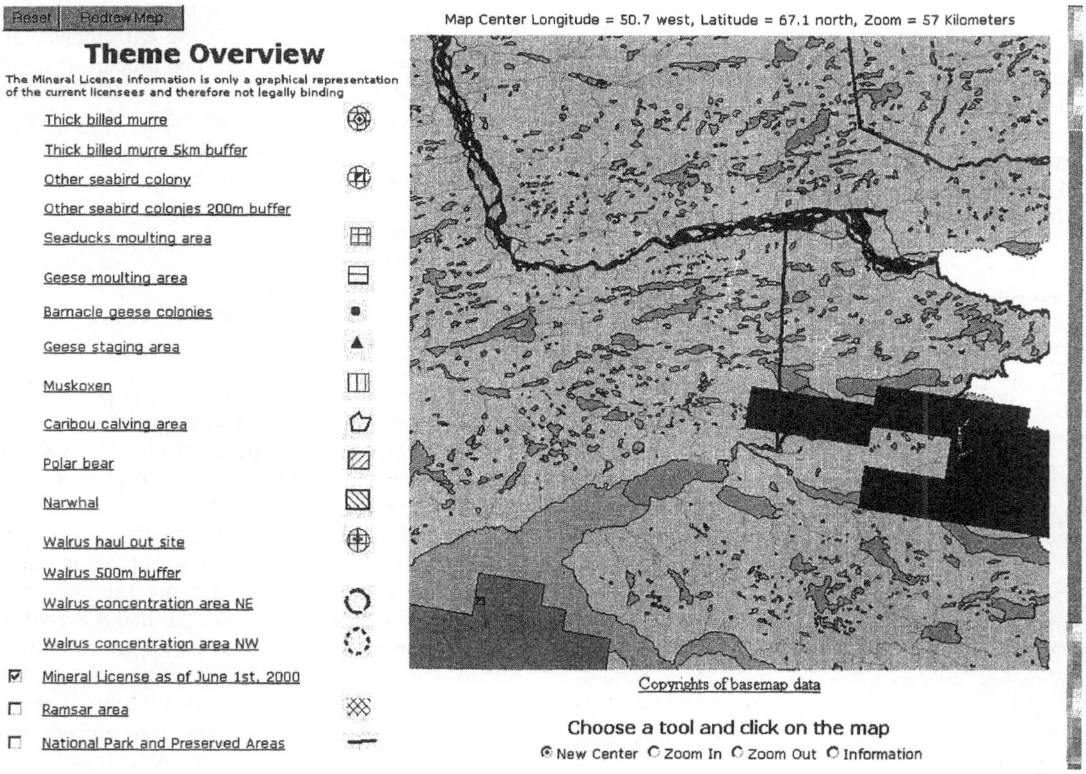

Figure 10. Web-page. License areas and calving range.

To help the mining companies to incorporate areas important to wildlife in their exploration planning, such areas and preserved areas have been mapped on the BMP homepage. Here, the companies can find their geological target areas, evaluate if any areas important to wildlife are situated within

REFERENCES

Aastrup, P. 2000. Responses of West Greenland caribou to the approach of humans on foot. Polar Research 19(1), 83-90.

Anon. 1993. *Sensitive habitats of the Porcupine caribou herd.* Report accepted by the International Porcupine caribou board from the Porcupine Caribou technical committee. 28 pp.

Ankney, C. D. & MacInnes, C. D. 1978. Nutrient reserves and reproductive performance of female Lesser Snow Geese. *AUK* 95:459-471.

BMP 1999. *Rules for field work and reporting regarding mineral resources (excluding hydrocarbons) in Greenland.* Government of Greenland, Bureau of Minerals and Petroleum, Greenland: 33 p.

Boertmann, D. & Glahder, C. 1999. *Grønlandske gåsebestande-en oversigt (Greenland goose populations-a review.* In Danish with English summary). Faglig rapport nr. 276. Danmarks Miljøundersøgelser, 59 p.

Cameron R.D., D.J. Reed, J.R. Dau & W T Smith 1992: Redistribution of calving caribou in response to oil field development on the arctic slope of Alaska. *Arctic 45(4): 338-342.*

Glahder, C. M. 1999a. Spring staging areas of the Greenland White-fronted Goose *Anser albifrons flavirostris* in West Greenland. *Arctic* 52 (3): 244-256.

Glahder, C. M. 1999b.Moulting Greenland White-fronted Geese: Distribution and concentrations in West Greenland: 118-142. In Glahder, C. M. *Sensitive areas and periods of the Greenland White-fronted Goose in West Greenland.* Ph.D. thesis. National Environmental Research Institute, Denmark, 142 p.

Glahder, C. M., Fox, A. D. & Walsh, A. J. 1999a. Satellite tracking of Greenland White-fronted Geese. *Dansk Orn. Foren. Tidsskr.* 93: 271-276.

Glahder, C. M., Nymand, J. & Petersen, M. K. 1999b. Feeding behaviour and habitat use of Greenland White-fronted Goose at a specific spring staging area: 67-94. In Glahder, C. M. *Sensitive areas and periods of the Greenland White-fronted Goose in West Greenland.* Ph.D. thesis. National Environmental Research Institute, Denmark, 142 p.

Glahder, C. M., Fox, A. D. & Walsh, A. J. (in prep.). Spring staging areas of the Greenland White-fronted Goose discovered by satellite tracking methods.

Hohman, W. L., Ankney, C. D. & Gordon, D. H. 1992. Ecology and management of postbreeding waterfowl: 128-189. In B. D. J. Batt, A. D. Afton, M. G. Anderson, C. D. Ankney, D. H. Johnson, J. A. Kadlec & G. L. Krapu (eds.), *Ecology and management of breeding waterfowl.* University of Minnesota Press, USA.

Madsen, J. & Mortensen, C. E. 1987. Habitat exploitation and interspecific competition of moulting geese in East Greenland. *IBIS* 129: 25-44.

Miller, M. W., Jensen, K. C., Grant, W. E. & Weller, M. W. 1994. A simulation model of helicopter disturbance of moulting Pacific black brant. *Ecological modelling* 73: 293-309.

Mosbech, A. & Glahder, C. 1991. Assessment of the impact of helicopter disturbance on moulting Pink-footed Geese (Anser brachyrhynchus) and Barnacle Geese (Branta leucopsis) in Jameson Land, Greenland. *Ardea* 79: 233-238.

Ogilvie, M. A., Boertmann, D., Cabot, D., Merne, O., Persival, S.M. & Sigfusson, A. Barnacle Goose (Branta leucopsis); Greenland: 246-256. In Madsen, J., Cracknell, G. & Fox, A.D. (eds.). *Goose populations of the Western Palearctic. A review of status and distribution.* Wetlands International Publ. No. 48. Wetlands International, the Netherlands & National Environmental Research Institute, Denmark, 344 p.

Raveling, D. G. & LeFebvre, E. A. 1967. Energy metabolism and theoretical flight range of birds. *Bird banding* 38:97-113.

Strandberg, B. 1997. Vegetation recovery following anthropogenic disturbances in Greenland. In: Crawford, R.M.M. (ed): Disturbance and recovery in arctic Lands.

Vistnes, I & Nellemann, C. 2000. Tap av kalvingsland som følge av forstyrrelse fra hyttefelt og kraftlinjer. *Reindriftsnytt 2/3:* 50-54.

Mining in the Arctic, Olsen, Lorentzen & Rendal (eds), © 2001 Swets & Zeitlinger, Lisse, ISBN 90 5809 177 5

Reactivity of covered sulfidic waste near Maarmorilik in West Greenland

B. Elberling
Institute of Geography, University of Copenhagen, Copenhagen, Denmark

G. Asmund
Department of Arctic Environment, National Environmental Research Institute, Risø, Denmark

T. Balic-Zunic
Institute of Geology, University of Copenhagen, Copenhagen, Denmark

T. Rohwerder & W. Sand
Institut für Allgemeine Botanik, University of Hamburg, Hamburg, Germany

ABSTRACT: Oxygen consumption was investigated in covered pyritic mine waste deposited on land near Maarmorilik in West Greenland. Pore gas concentrations of oxygen and carbon dioxide were measured in waste and in cover material. Samples of mine waste were collected for calorimetric evaluation of site-specific and temperature-dependent sulfide oxidation. To evaluate the mineral composition of waste material, additional samples were taken for quantitative X-ray diffractometry. Results show that reactive pyrite grains exist below the cover material and that pyrite oxidation results in an oxygen gradient across the cover. The rate of oxygen consumption was generally low, which is consistent with the observed pH values between 8 and 9. However, a sample from a spill of pyritic sediment near the covered area indicates that sulfide oxidation occurs at high rates, in accordance with the high concentrations of heavy metals observed in water draining from the area.

1 INTRODUCTION

Low annual temperatures and permafrost may offer several opportunities for limiting acid mine drainage (AMD) from sulfidic material deposited on land in the Arctic. Frozen conditions have the potential to reduce the chemical reactions that produce AMD and the subsequent movement of contaminants. However, due to annual thawing an active layer will develop, which will be infiltrated by water and atmospheric oxygen. As a consequence, an oxidation will be initiated if reactive sulfidic material is present. The thickness of the active layer varies depending on local ground conditions and weather. Covering the sulfidic waste with an inert material may hinder sulfide oxidation, concurrent acid generation and heavy metal release.

Pyrite is among the most common sulfide minerals which oxidize in the presence of water and oxygen. The complete oxidation of the iron and sulfur moieties of pyrite in an alkaline environment can be written as:

$$FeS_{2(s)} + 3.75O_{2(g)} + 0.5H_2O \rightarrow 2SO_4^{2-} + Fe^{3+} + H^+ \quad (1)$$

The influence of low temperatures on the overall reaction rate of pyrite oxidation can be described by the Arrhenius equation (Nicholson et al., 1988):

$$\ln (k_1/k_2) = E_a (T_1 - T_2) / (RT_1T_2) \quad (2)$$

where k_1 and k_2 are reaction rates at temperatures T_1 and T_2, E_a is the activation energy of the reaction and R is the gas constant. Values of E_a for pyrite oxidation reported in the literature range from 39 to 88 kJ/mol FeS_2 under a variety of experimental conditions (Nicholson et al., 1988). Although the effects of temperature on sulfide oxidation rates can be complex, particularly under field conditions, several studies confirm that the relatively simple relationship given by Equation 2 is a valid approximation of biotic and abiotic reaction rates (Ahonen & Tuovinen 1992, Dawson & Morin 1996). Dawson & Morin (1996) reviewed reaction rates related to *Thiobacillus ferrooxidans* and concluded that the reaction rate generally decreases from maximum rates around 30°C to rates several orders of magnitude lower at temperatures near 0°C. However, *Thiobacillus* bacteria have been found to adapt to very low temperatures. Kalin (1987) reported the presence of psychrophilic (growth at or around 0°C with a maximum temperature for growth < 21°C) *Thiobacillus* near Nanisivik Mine in Arctic Canada. Langdahl & Ingvorsen (1997) studied *Thiobacillus* bacteria from North Greenland and reported maximum bacterial pyrite oxidation rates at approximately 20-25°C and 30% of the maximum rate at 0°C. Although the occurrence of sulfide-oxidizing bacteria in the High Arctic has been reported, the actual activity and importance of bacteria for sulfide oxidation in cold climates have not received much attention.

Additional environmental controls on the overall

oxidation of sulfide minerals have been extensively studied. Nicholson et al. (1988), reporting pyrite oxidation to be a first-order reaction with respect to specific surface area, emphasized the importance of pyrite grain size. Higher oxygen concentrations also increase the oxidation rate but the effect decreases as atmospheric oxygen concentrations are reached. At low pH values the kinetics of Fe^{2+} oxidation and bacterial catalysis seem to increase the rate of pyrite oxidation substantially (Nicholson et al. 1988). Although moderately acidophilic or neutrophilic thiobacilli (pH optima of 5-8) have been found in ore samples with neutral pH values (Schippers et al. 1996), biotic rates are slow at neutral pH values. To eliminate biological controls on sulfide oxidation, carbonate minerals or rocks are often mixed with sulfidic waste or used as a cover material to buffer acid generation, thereby maintaining neutral pH values in sulfidic waste material.

Despite their relevance for control strategies to limit AMD from sulfidic waste deposited in the Arctic, only few field studies have evaluated the actual reaction kinetic and potential AMD in covered pyritic waste deposited in the Arctic. Near the former Black Angel Mine (Maarmorilik) in West Greenland, pyrite is the dominant sulfide mineral in waste material which has been covered since 1991 by 40-70 cm of sediments consisting primarily of carbonate minerals. It has been shown that the waste material contains heavy metals which may be released and mobilized in the case of acid mine drainage (Asmund 1992). This paper aims to quantify the oxygen uptake and depth-dependent sulfide oxidation in covered waste and to discuss the environmental factors controlling the rates observed and the potential release of heavy metals.

2 MATERIALS AND METHODS

2.1 *Study area*

Maarmorilik is located in the inner part of the Uummannaq fiord system in West Greenland (71°07N; 51°15W). The climate is arctic with an annual mean temperature of -8°C and is influenced by strong winds from the inland ice, which result in sparse precipitation (<120 mm). A massive sulfide ore deposit with approximately 24-30% pyrite (FeS_2), 14-15 % Zn and 4% Pb was mined and milled from 1973 to 1990. Over that period, the mine processed about 11 million tonnes of ore. The mine tailings produced were discharged into a small adjacent fiord, Affarlikassaa which is partly separated from the outer fiord by a small sill. Waste rocks have been dumped outside the mine since 1973. Metal contamination beyond the marine disposal area has long been documented (e.g., Asmund 1992) and several sources of heavy metal pollution have been identified, with tailings, waste rock, dust and concentrate being the most significant. As part of the mine closure, sulfidic waste material near the tidal zone was removed, and partly dumped in Affarlikassaa on top of the tailings (320,000 tones) and partly placed on land (80,000 tones). Subsequently, the waste material on land was covered by 40-70 cm of carbonate rocks. The heavy metal composition of the waste have been reported to be 0.89±0.2 %Pb, 2.7±0.4 %Zn and 128.8±24 mg/kg Cd based on four replicates (Asmund 1992)

2.2 *Field measurements*

Field work completed at the end of August 1999 included vertical gas profiling and sampling of cover material and underlying waste material. Oxygen (O_2) and carbon dioxide (CO_2) concentrations were measured at 10±0.5 cm depth intervals by a modified technique described by Elberling & Nicholson (1996). Pore gas from a specific depth was withdrawn by connecting a stainless steel tube (Dräger) to a portable Carbon Dioxide and Oxygen Analyzer PCO4 (Gas Data Ltd). The analyzer has previously been investigated for its reliability under variable field conditions and was found to provide measurements to an accuracy of 0.05% O_2 and 0.005% CO_2. The total gas volume of the sampling equipment was less than 70 ml. Based on several O_2 profiles measured across the inert cover material, the flux of O_2 was calculated by Fick's first law:

$$F = D_{eff}\, dC/dx \qquad (3)$$

where F is the diffusive flux, C is the O_2 concentration at depth x, and D_{eff} is the effective molecular diffusion coefficient. D_{eff} was estimated on the basis of the free diffusion in air, the soil porosity and the water content (Millington 1959). The O_2 flux across the cover corresponds to the total O_2 consumption in a vertical profile below the cover, assuming that the oxygen gradient reflects steady state conditions and that one-dimensional oxygen diffusion is the major transport mechanism responsible for subsurface gas transport. This issue has previously been discussed (e.g., Elberling & Nicholson 1996).

Depth-specific samples of cover and waste material were collected at 10-cm intervals. Bulk samples as well as a sieved fraction (< 2 mm) were collected under sterile conditions and stored in cups of 100 cm^3. Sampling cups were only half filled to allow oxidation. For comparison naturally exposed waste material was sampled from the area surrounding the covered waste. Samples were kept dark at between 4 and 6°C until laboratory processing within a week.

Pore solutions were collected by Prenart Super Quartz® suction probes installed horizontally in the active layer downstream of the covered waste area. The suction probes, made of PTFE (Teflon), had a pore radius of 2 microns and a hydraulic conductivity of 3.31×10^{-7} cms^{-1}. They were installed in August 1999 while the subsequent pore water extraction and sampling took place in 2000. Teflon tubes were used to connect the probes to 250-ml sampling bottles placed on the soil surface. To ensure high spatial and temporal resolution of the sampling, small volumes (about 100 ml) were sampled over 24 hours with a suction of approximately 0.3 atm.

2.3 *Laboratory measurements*

A quantitative mineralogical description of samples was based on visual comparisons of actual X-ray powder diffraction diagrams and theoretical diagrams calculated from the known crystal structures of minerals. For further chemical analysis 2 g of each sample was added to 20 ml of a 0.9% NaCl solution in 100-ml flasks. For extraction of soluble compounds the flasks were shaken on a rotary shaker for 2 hours at 28°C. After centrifugation for 10 min (Heraeus, 9,000 rpm), sulfate was determined by ion chromatography (Dionex, model DX 500). pH measurements were made in a suspension of 5 g sample and 20 ml water using an Ingold electrode. The water content was determined as weight loss after 48 hours at 60°C. Other measurements included grain size distribution and porosity of samples. Pore water samples were analyzed for total Zn, Pb, and Cd concentrations by anodic stripping voltommetry.

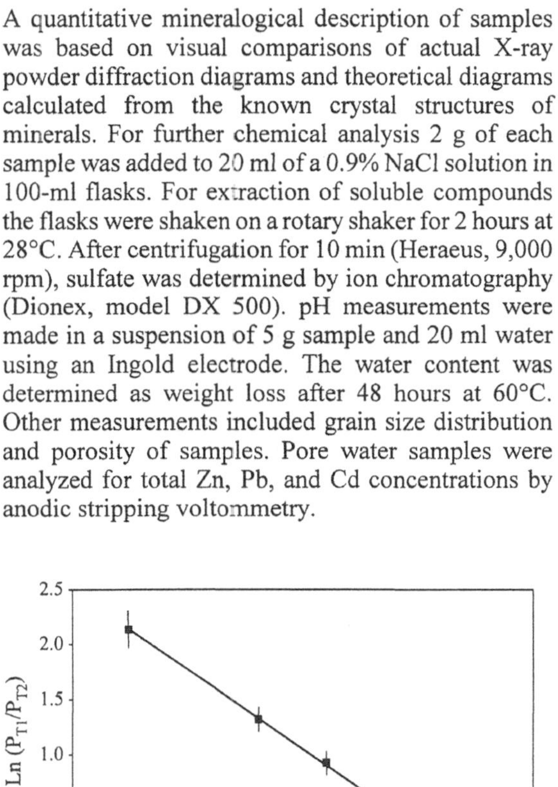

Figure 1. An Arrhenius plot showing normalized heat output to 5°C (P_{T2}) for measurements made at 5, 10, 15, 20, and 30°C (n = 32). The line represents linear regression (r^2 > 0.99) and bars show one standard deviation.

The pyrite oxidation rates of samples were determined as the heat output of the samples at 20°C measured using a thermal activity monitor (type 2277, Thermometric, Sweden, or C3-Analysentechnik, Germany) equipped with ampule cylinders (20 ml twin, type 2230). Glass ampules containing 15 g of material were inserted in the measuring cylinders and, after thermal equilibration, the heat output resulting from the microbial degradation of pyrite was recorded, as previously described (Sand et al. 1993). Additional measurements were made at 5, 10, 15, 20 and 30°C in order to establish an Arrhenius relationship (Eq. 2). Given the generally low heat output at low temperatures, measurements were not made below 5°C. Only one sample had enough heat output to produce reliable rates at 5°C. To measure the heat output caused by pure chemical oxidation, the bacteria were exterminated by chloroform. After 2 hours the chloroform was removed by vacuum evaporation and heat output was measured again.

3 RESULTS AND DISCUSSION

3.1 *Mine waste characteristics*

Waste rock from the study area is a sand and gravel material with about 50 vol% carbonate represented as approximately equal volumes of calcite and dolomite. The fraction finer than 2 mm (approximately 50% of total waste) contains about 15 and 40% pyrite, 1-2% sphalerite and less than 1% galena. In waste material coarser than 2 mm, the total sulfide content is less than 1%. The rest is represented by silicate minerals and quartz. The cover consists of more than 90% carbonate minerals and small amounts of quarts, albite and other silicate minerals. The pH values of the waste material were between 8.1 and 9 and the environment can therefore be characterized as well buffered. The pH values in naturally exposed samples were lower but above 7.

3.2 *Temperature-dependent activity*

The relative heat output versus temperature for one sample is shown in an Arrhenius plot (Fig. 1) with a correlation of $r^2 > 0.99$ (n = 32) and Ea = 59.9 ±1.1 kJ/mol. For most samples the heat output at low temperatures (≤10°C) was close to the detection limit. Consequently, the temperature relationship established is based on one sample. It is assumed that the temperature-dependent activity for this sample is representative for sulfide oxidizing in the area as the observed Arrhenius relationship is almost identical with results from a detailed calorimetric study on pyritic waste from Nanisivik Mine (Elberling et al. 2000). Because of the degree of correlation observed, the Arrhenius equation is used in the following to convert measured activities at 20°C to actual activities by taking observed ground temperatures as low as 0°C into account.

3.3 *Total and depth-dependent reactivity*

Ground temperatures at the field site were approximately 8°C near the surface, 5°C below the cover, and 0°C at the permafrost level at a depth of approximately 1.4 m in August (Fig. 2A). The temperature profile represents only a relatively short period of time as the active layer develops throughout the summer and freezes up during September and October. The temperature profile shown is assumed to represent the warmest situation on an annual basis.

A drop in the oxygen gas concentration across the inert cover was observed at all sites (Fig. 2B). Below the cover the O_2 concentration decreases further as a result of the O_2 uptake by sulfide oxidation. As pyrite oxidation (Eq. 1) is considered the main oxygen-consuming process the O_2 gradient across the cover reflects the total O_2 uptake below. The total O_2 consumption in the vertical profile can be estimated to be 0.1 moles m^{-2} d^{-1} (±0.05) based on the observed O_2 gradients (n = 8) across the cover and a calculated average D_{eff} of 1.86×10^{-6} m^2s^{-1}. Due to the high concentrations of O_2 throughout the active layer, the availability of O_2 is not limiting the overall O_2 consumption in the upper 80 cm. The subsurface O_2 uptake corresponds to a CO_2 production (Fig. 2B) probably resulting from carbonate dissolution, which can be considered the natural buffering of acid produced by pyrite oxidation.

Figure 2C shows calorimetric data from covered and well-drained waste material (< 2 mm) based on the average measured activity at atmospheric O_2 concentrations at 20°C. The profile reveals an increasing activity just below the cover and fairly stable activities at lower depths. This trend is consistent with changing concentrations of sulfate, one of the dominant reaction products from pyrite oxidation. The low activity measured at 35-cm depth corresponds in this profile to a sample with only traces of pyrite and the sample which can be considered part of the cover and only contains traces of pyrite. Other variations in activity at greater depth could not be correlated with the pyrite content, although variations in specific surface area and reactivity of single pyrite grains are likely to influence the overall activity.

In order to compare heat output with the observed total oxygen uptake across the cover material, the total heat production observed is converted to a total pyrite oxidation (moles m^{-2} d^{-1}) as previous shown (Elberling et al. 2000). First heat outputs measured at 20°C on individual samples have been converted to a rate representative for the temperature measured during sampling by using the Arrhenius relationship (Fig. 1). The total heat output within the profile has then been estimated as the depth-integrated heat output and further converted to a pyrite oxidation rate, using a reaction energy value for pyrite oxidation of 1546 kJ/mol, given for biological pyrite oxidation by calo-

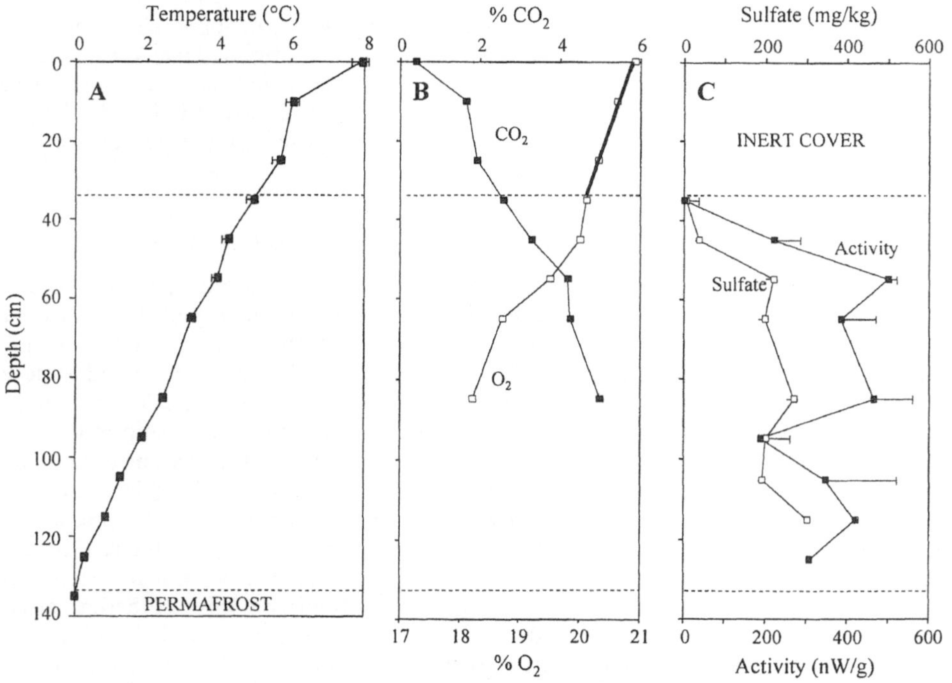

Figure 2. Depth-dependent ground temperatures, pores gas composition, solid sulfate content and total heat output at 20°C from oxidizing pyritic waste. The thick solid line across the inert cover represents linear regression of oxygen concentrations ($r^2 > 0.99$) and bars represent standard deviations.

rimetry (Rohwerder et al. 1998). The final conversion to total oxygen consumption within the profile per surface area takes into account the depth-dependent density of the tailings (between 2.9 and 3.1 g/cm^3), the porosity (38%), the water content (8-10%), and the stoichiometric conversion of pyrite oxidation by oxygen (Eq. 1). This final calculation gives a total oxygen uptake of 0.08 moles $m^{-2} d^{-1}$. This is lower than the average estimate based on the oxygen gradient method, but within the natural variation given by ± 1 standard deviation. The agreement in the methods applied is consistent with a laboratory study showing that a calorimetrically determined reaction energy value for chemical and bacterial pyrite oxidation agrees with the theoretical value according to Equation 1 (Rohwerder et al. 1998). Thus, the amount of oxygen consumed can be determined by calorimetric measurements. Also, results from laboratory measurements of calorimetric pyrite oxidation activity have previously agreed well with *in-situ* oxygen uptake rates measured with closed chambers (Elberling et al. 2000). However, the measurement of pyrite oxidation activity at atmospheric oxygen concentrations may have caused deviation and resulting rates may therefore not be representative of *in-situ* conditions. However, in this case oxygen is not considered limiting for the near-surface sulfide oxidation rates in the field and atmospheric concentrations during laboratory experiments are therefore not considered critical for the overall heat output.

3.4 *AMD and heavy metal release*

Chemical analysis of pore water collected in 2000 by suction probes downstream of the covered waste area are shown in Table 1. The high concentrations of metals in drainage water from land reveal that AMD from sulfide oxidation on land contributes to the pollution of metals in Affarlikassaa. However, metal concentrations in Affarlikassaa probably stem from near-bottom dissolution and oxidation processes rather than release of metals from land as metal concentrations increase with depths in the fiord.

Table 1. Dissolved metals (µg/kg) in pore water collected on land and in depth-specific water samples from Affarlikassaa.

	Zn	Cd	Pb
Fiord 0-20 m below surface	0.5	0.01	0.1
Fiord 30-50 m below surface	15	0.11	0.3
Fiord 5 cm above bottom	20	0.12	0.3
Drainage water on land # 1	10,250	53.4	32.6
Drainage water on land # 2	2,660	11.8	932.9

Both drainage water samples contained high but varying concentrations of metals, a difference possibly caused by the different local sources of the metals. The composition of the waste material is fairly heterogeneous, which may explain some of the variation observed. These preliminary studies do not provide any basis for determine the total amount of metal released nor whether the metals are released within the covered waste or from other sources within the drainage area. However, it seems likely that ongoing processes are responsible for the observed metal release.

4 ENVIRONMENTAL APPLICATIONS AND CONCLUSIONS

Variations in heat output from depth-specific samples versus pH are shown in Figure 3. A decreasing heat output can be observed with increasing pH but the relationship is not significant (r^2 = 0.47). The percentage of chemical oxidation was 96% for the sample with the lowest pH, confirming the limited bacterial influence compared with an acidic environment. The trend observed in Figure 3 may results from the importance of pH for the percentage of bacterial catalysis and thereby for the overall rate of sulfide oxidation. Calorimetric data from a study based on well buffered sulfidic tailings deposited near Nanisivik Mine reveal that microbial pyrite oxidation at pH between 6 and 7 accounts for about one-third of the total activity (Elberling et al. 2000). Additional measurements made on non-buffered tailings in Nanisivik (pH below 3) were found to be 100% biologically controlled.

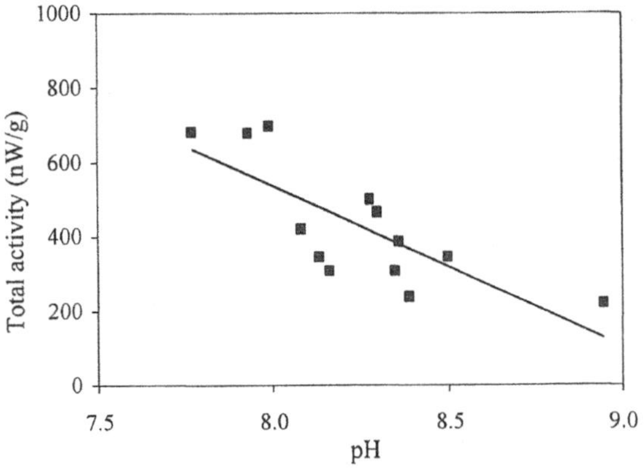

Figure 3. Total heat output versus pH of sulfidic waste covered by carbonate rocks. The solid line represents linear regression (r^2 = 0.47).

The Maarmorilik observations indicate that the combination of low temperatures, a high pH and the relatively coarse nature of sulfide grains (compared to tailings) is important for the overall low oxygen uptake rate and almost negligible biotic catalysis in covered waste. This study presents only a snapshot of the actual oxygen uptake in the covered waste; however, the observed rate probably represents the maximum oxygen uptake on an annual basis. This implies that the AMD generation in the coarse carbonate-covered sulfidic waste in Maarmorilik is low. However, the high heat output measured in sulfidic material from an uncovered site reveals that high sulfide oxidation rates occurs in the area and that this rate is similar to that observed for tailings deposited on land near Nanisivik Mine (Elberling et al. 2000). This uncovered sulfidic material in Maarmorilik is probably a spill of tailings from the mining period. The distribution of such reactive material in the area has not been investigated, but only small areas covered with this material will produce AMD, a factor which is far more important for the overall metal contamination in the area than processes occurring in the covered waste.

In tailing impoundments, the diffusion of oxygen gas is often considered a controlling parameter for the overall oxidation (e.g., Pantelis & Ritchie 1991, Elberling et al. 1994). However, given the low rates and high effective diffusion coefficient in the covered waste in Maarmorilik, oxygen cannot be considered rate limiting at least not in the upper 80 cm of the waste material. Therefore, the Maarmorilik observations cannot be used as a general verification of the environmental effects of covering reactive sulfidic material in the High Arctic with carbonate rocks.

ACKNOWLEDGMENTS

This study was funded by the Danish Cooperation for Environment in the Arctic (DANCEA), Danish Environmental Protection Agency, Ministry of Environment and Energy (Denmark). The authors are solely responsible for the content of this paper. The study was further supported by the National Environmental Research Institute and Institute of Geography (University of Copenhagen).

REFERENCES

Ahonen, L. & O.H. Tuovinen 1992. Bacterial oxidation of sulfide minerals in column leaching experiments at suboptimal temperatures. *Appl. Environ. Microbiol.* 58: 600-606.

Asmund, G., 1992. Lead and zinc pollution for marine dumping of waste rock from lead/zinc mining. In: *Mining in the Arctic*, Bandopadhyay & Nelson (eds). Balkema, Rotterdam, 105-112.

Dawson, R.F. & K.A. Morin 1996. Acid mine drainage in permafrost regions: issues, control strategies and research requirements. MEND Project 1.61.2. *Departments of Indian and Northern Affairs Canada.* Ottawa, Ontario, Canada.

Elberling, B., R.V. Nicholson & J.M. Scharer 1994. A combined kinetic and diffusion model for pyrite oxidation in tailings: a change in controls with time. *J. of Hydrol.* 157: 47-60.

Elberling, B. & R.V. Nicholson 1996. Field determination of sulphide oxidation rates in mine tailings. *Water Resour. Res.* 32(6): 1773-1784.

Elberling, B., A. Schippers & W. Sand 2000. Bacterial and chemical oxidation of pyritic mine tailings at low temperatures. *J Cont. Hydrol.* 41: 225-238.

Kalin, M., 1987. Ecological engineering for gold and base metal mining operation in the North West Territories. In: *Northern Affairs Program, Department of Indian Affairs and Northern Development*, Environmental Studies no. 59, pp. 1-64.

Langdahl, B.R. & K. Ingvorsen 1997. Temperature characteristics of iron solubilisation and ^{14}C assimilation in natural exposed ore materiel at Citronen Fjord, North Greenland (83°N). *FEMS Microbial. Ecology* 23: 275-283.

Millington, J. 1959. Gas diffusion in porous media. *Science* 130: 100-102.

Nicholson, R.V., R.W. Gillham & E.J. Reardon 1988. Pyrite oxidation in carbonate-buffered solution: 1 experimental kinetics. *Geochim. Cosmochim. Acta.* 52: 1077-1085.

Pantelis, G. & A.I.M. Ritchie 1991. Macroscopic transport mechanisms as a rate limiting factor in dump leaching of pyritic ores. *Appl. Mat. Model.* 15: 136-143.

Rohwerder, T., A. Schippers & W. Sand 1998. Determination of reaction energy values for biological pyrite oxidation by calorimetry. *Thermochimica Acta.* 309: 79-85.

Sand, W., R. Hallmann, K. Rohde, B. Sobotke & S. Wentzien 1993. Controlled microbiological in-situ stope leaching of a sulphidic ore. *Appl. Microbiol. Biotechnol.* 40: 421-426.

Schippers, A., P.G. Jozsa & W. Sand 1996. Sulfur chemistry in bacterial leaching of pyrite. *Appl. Environ. Microbiol.* 62: 3424-3431.

Mining in the Arctic, Olsen, Lorentzen & Rendal (eds), © 2001 Swets & Zeitlinger, Lisse, ISBN 90 5809 177 5

Mining in environmentally sensitive areas – the decision maker's dilemma.

Chris Norman
Chief Planning Officer (Development Co-ordination) West Lothian Council, Scotland

ABSTRACT: Mining and quarrying proposals are perhaps the most controversial types of land use change that planning authorities must decide upon. Decision makers have to carry out the difficult balancing act between, on the one hand, allowing for the extraction of minerals that are needed for society's growth, and, on the other protecting the environment for the sake of local communities, the landscape and the environment generally. This paper sets out how attempts are made in Scotland to reconcile these competing demands. Specific reference is made to the recent decision of the Scottish government to refuse planning permission for what would have been Europe's largest quarry in the remote Western Isles. Reference is also made to best practice methods for mineral working and the steps that regulatory bodies and industry can adopt to ensure that, where permitted, mineral working is carried out in an environmentally sustainable manner.

1 INTRODUCTION

Minerals are a valuable natural resource essential for the social and economic well being of a nation's economy. Society's insatiable demand for development, prosperity, and full employment means, in turn, an ever escalating quest for natural resources that are able to meet the demands of the market place. As minerals that are capable of easier working are depleted, industry's attention is becoming focused on new areas for mineral extraction, where geology has provided the raw materials, but a judgment must be made on whether their extraction may proceed because of broader environmental reasons.

But minerals can only be worked where they are found.

Competing interests may well prevail between, on the one hand, the winning and working of minerals and, on the other, the need to protect the environment and host communities from what is frequently seen as a threat to local well being and the environment generally.

We live in the era of sustainable development; environmentalism is on the ascendancy and, generally, peoples' tolerances of unwelcome development is diminishing rapidly. In Scotland, new plans for mining and quarrying are often the most controversial types of development proposals that planning authorities must deal with and some proposals can now attract well over 1000 objections.

For those involved in the decision making process and who have to determine whether a mining or quarrying project should proceed, a delicate balancing act must be performed.

In this paper I set out what I see are the principle considerations that decision makers must take into account when assessing proposals for mining in environmentally sensitive areas. Reference will be made to best practice advice on mineral extraction, as set out by the Scottish government.

By way of example, I will describe the background to the recent decision of the Scottish government to refuse planning permission for Europe's largest quarry in the remote and beautiful Western Isles of Scotland. Here, despite local employment benefits, the environmental impact of the development on the scenery of a protected area caused the case to be dismissed, after Scotland's longest ever public inquiry.

Thereafter, I will describe ways in which the industry and regulatory bodies can assist each other in ensuring that, where mining and quarrying is permitted, it can proceed in an environmentally sustainable manner.

2 MINERALS PLANNING SCOTLAND

In Scotland, the determination, monitoring and enforcement of planning control over mineral working is carried out by 32 local authorities. Our northernmost planning authority, Shetland Islands Council lies north of Cape Farewell, Greenland. Under our recently devolved system of government, ultimate decision making is vested in the Scottish Ministers in Edinburgh. Planning law means that planning permission is needed for all types of mineral development, with the exception of some exploration work and limited ancillary mining development.

In making a decision on a planning application, a planning authority must take into account the development plan for the area, matters of government policy, replies from statutory consultees and representations from other interested parties, including affected local communities. Mineral planning applications are normally accompanied by an environmental impact statement prepared by the developer and assessed by the planning authority. A recommendation is then made by planning officials and a decision is made by elected members who sit on an authority's planning committee.

In Scotland, government guidance on mineral working is set out in National Planning Policy Guideline 4 (NPPG 4) 'Land for Mineral Working' published by the Scottish Office in 1994. The document sets out the range of matters to be taken into account in assessing mineral working proposals. The document was augmented in March 1999 with the publication of a second policy document, NPPG 16, concerning the surface extraction of coal and related minerals.

Planning authorities are provided with further advice from Government on detailed aspects of mineral working through a series of Planning Advice Notes (PANs).

3 ENVIRONMENTALLY SENSITIVE AREAS

In Scotland there is a graded classification of sensitivity for environmentally important areas. The highest tier of designation includes areas recognized internationally for their flora or fauna.

Such sites are classified under the European Union Directive for the Conservation of Wild Birds (Special Protection Areas, SPAs); the Directive on Flora, Fauna and Habitats (Special Areas of Conservation, SACs), or if a wetland is of international importance, under the Ramsar Convention 1975.

Mineral working in these areas will only be permitted in the most exceptional circumstances where the development will not adversely affect the safeguarded habitat or species or there are overriding nationally important reasons in the national interest for the development to take place.

The second tier of environmental sensitivity in Scotland includes those areas of national importance and where the environmental assets of the designated area are to be protected wherever possible. Here, mineral extraction is not prohibited but it must be reconciled with conservation interests. Environmental designations of national importance are National Scenic Areas (NSAs), National Nature Reserves (NNRs), Sites of Special Scientific Interest (SSSIs), Environmentally Sensitive Areas (ESAs) and national and regional parks.

Mineral extraction in this second tier should only be permitted where the objectives of designation and the overall integrity of the area will not be compromised; or any significant adverse effects on the qualities for which the area has been designated are clearly outweighed by social and economic benefits of national importance. In this case, national benefits could include employment creation and balance of payments considerations.

In addition to areas recognized for their natural heritage importance, it is my view that a human presence in the vicinity of a mineral working proposal means that such an area also becomes environmentally sensitive.

Government guidance seeks to address this in the March 1999 publication of NPPG 16, prepared in response to increasing pressures for opencast coal mining in central Scotland.

Here, guidance acknowledges that the identification of towns and villages as communities is self evident. Although, only exceptionally, are groups of less than 10 houses likely to constitute a community, it is planning authorities that are required to decide on the most appropriate identification of 'communities'. Indeed I believe it is important for individual residents to expect protection from the planning system and it is somewhat false to regard individual or groups of 3 or 4 houses as being unworthy of such safeguarding.

NPPG 16 requires that particular attention needs to be given to separation distances between proposed sites and adjacent housing. As a general rule, working faces within 500m of a community are regarded as likely to pose a threat to the amenity and are likely to be unacceptable.

In the interests of protecting communities and the local environment from the consequences of mineral working, and in particular opencast or strip mining, government policy requires the following tests to be applied.

Specifically, the proposal must be environmentally acceptable, taking into account the use of effective regulation to offset or mitigate any adverse impacts. If it is not, there must be local or community benefits related to the proposal which significantly outweigh any material risk of disturbance or environmental damage.

4 PUBLIC CONCERN AND MINERAL PLANNING

The reaction of the public to a minerals planning application is a factor that must be taken into account in the decision making process. The planning and environmental impact assessment process places a statutory obligation on regulatory bodies to publicise details not only of a proposal, but any additional environmental information received and the reasons why a proposal is to be granted or refused development consent.

It is clear that the public have a right to environmental information and a planning authority must have regard to the views of objectors. Any summary dismissal of those points of objection could cause a legal challenge to a decision under environmental or, more recently, human rights legislation. Indeed, having fulfilled a statutory responsibility for requesting views, the public have a legitimate expectation that the decision maker will take into account their views and uphold them in making a decision. Moreover, the concerns of individuals do not have to be scientifically proven to be relevant to the decision making process; legal case law has confirmed that reasonably engendered fear which justifies an emotional response to a project can be enough to warrant the refusal of planning permission.

Ultimately, the weight to be attached to public concern expressed about environmental matters is a matter of judgment and will depend on issues such as evidence, causation and remoteness to the development proposal.

5 ENVIRONMENTAL IMPACT ASSESSMENT

As in other parts of the European Union, in Scotland European law almost always takes precedence over national law. In March 1999, the amended European Union Directive 97/11/EC invoked much wider ranging environmental impact assessment regulations. These amended regulations require EIA to be carried out on all mineral projects extending over 25ha and, additionally, to many smaller schemes. Each environmental impact statement is required to contain a broad range of environmental information and to describe the effects that the development will have on the environment.

There can be little doubt that the environmental impact statement allows the decision maker to carry-out a systematic assessment of the environmental consequences of mining proposals. It is an inevitable consequence of the Scottish system of EIA that the process can be used by developers to justify a proposal, whatever the environmental consequences. Of the 30 or so environmental statements that I have assessed, I have yet to see one that concludes that development should not proceed because of it's environmental effects. Indeed, I tend to see an environmental statement as a base line and it is the planning authority's function to either prove or disprove the assumptions that it contains.

There are good environmental statements, and, quite frankly, there are some that are not worth the paper they are written on, even though they make an attempt to defend the indefensible in planning and environmental terms.

An environmental statement needs to be honest, frank and focused; many are simply justifications for development to proceed, whatever the environmental costs

6 KEY FACTORS IN THE DECISION MAKING PROCESS.

Having very briefly described an overview of the Scottish planning system, it's relationship with environmentally sensitive areas, the environmental impact assessment framework and the role of public concern, I now wish to turn to the main determining issues of a minerals planning application.

Landscape and visual impact
There is no escaping the fact that mineral working is visible and in all probability will be intrusive into the landscape. Landscape features can be removed or altered for ever and an industrial activity will be introduced into what may well be an attractive landscape. Arguably, landscape change is one of the most controversial aspects of mining and quarrying. Government guidance in Scotland confirms that landscape considerations and visual intrusion are thus major issues at the planning application stage.

Landscape impact assumes an even greater importance if the development is proposed in an area the subject of a landscape designation, either nationally (national scenic areas) or locally (areas of great landscape value). In national scenic areas, which cover around 12.7% of Scotland's surface area, the overall integrity and objectives of designation must not be compromised, or there must be social or economic benefits that clearly outweigh any significant adverse effects. Similar, but less stringent tests apply to areas of local landscape importance.

A difficulty faced by mineral planners is the need to quantify the qualitative aspects of landscape attributes and decide what weight must be given to landscape matters. In Scotland, a national programme of landscape character assessment is nearing completion. Here a scientific methodology has been used to classify and describe the landscape. The capacity of the landscape to absorb change and visual intrusion can then be determined. The value of the landscape in experiential terms is set out and special characteristics of an area are described.

The tool is at least a start for planners; however the capacity of the landscape to absorb the development is also dependent on the scale of the proposal, the extent of sensitive visual receptors and the importance of the landscape in national and local terms.

Particular care is needed in the design and internal layout of a minerals project if it is to proceed. Screening mounds that are unnatural and engineered in appearance in their form can often look as incongruous as the mining works themselves. Careful design and layout are important mitigating factors

Sensory environmental effects of noise, dust and blasting

Noise generated by mineral workings is often, in my opinion, the most frequent cause of complaint. Particularly in quieter rural areas, the intrusive and alien noise generated by mechanical plant can disturb people in their homes and those enjoying passive outdoor recreation. Wildlife, especially in the breeding season, can also be disturbed by noise from mineral workings.

In determining a mineral planning application, it is of great importance to understand the background, pre-existing noise levels at a site. Noise predictions are necessary, based upon estimates of each item of plant and taking into account mitigation measures and meteorological patterns.

The background L_{A} 90 noise level (i.e. that occurring for 90% of the time) should not generally be exceeded by more than 10dB or complaints can otherwise be expected. Alternatively, Planning Advice Note 50 (PAN 50) suggests levels of noise that are appropriate at different times of the day at noise sensitive receptors. During the daytime, noise levels should not generally exceed 55dB L_{Aeq} 1hr; the night time noise limit is set at 42 dB $L_{A\ eq}$ 1 hr. It should be noted that draft guidance applicable in England is suggesting considerably more stringent noise level standards.

Nevertheless, I am a firm believer in examining the relationship between pre-existing and introduced noise on a mining site. Proper noise monitoring is important to ensure that developers comply with planning conditions imposed upon them.

Dust is acknowledged by Government as being a major environmental concern and it is recognised that residents living up to 1 km from a mine can be affected by dust fallout. Dust can tarnish vegetation around a mine or quarry with adverse effects on the appearance of an area and for biodiversity generally. Bryophytes and lichens are especially susceptible to fugitive dust from surface mineral workings. Because it is closely related to climatic trends, dry windy weather is more likely to cause dust problems.

Extensive public concern has been expressed recently about the effects of airborne particulate matter from mines on the respiratory health of residents, and especially on children and other vulnerable members of society. Draft Government guidance in Scotland is now calling for an 'air quality assessment' to be carried out for sites within 1km of settlements when PM_{10} particulate matter is predicted to exceed current national air quality standards.

The regulation of dust from mines in Scotland tends to rely on best practice techniques rather than the setting of absolute standards; a wide range of dust suppression methods are expected to be used by mining companies. A formal environmental management system is seen as essential for mineral sites where dust is identified as an issue.

Blasting can produce an invasive effect that can impact on people's homes and the enjoyment of their amenity. It can also startle wildlife, again especially in the breeding season. This difficulty is exacerbated by the fact that humans are particularly sensitive to vibration and can perceive the effects of blasting even below levels that can cause cosmetic damage in buildings.

While standards for ground vibration are becoming more vigorous, it remains important to ensure that not only is damage to property avoided, but perceived fear and disamenity are addressed. Levels of peak particle velocity (ppv) at 12mm/second are extremely high; even ppv's of 6mm/second can have adverse effects when felt in a house. However, this range tends to be the standards to which industry wishes to operate. The human response to blasting dictates that much lower levels of ground vibration are necessary if the mineral extraction industry is to continue to be a good neighbour.

The water environment

Mineral workings can have effects on both surface and groundwater. Specifically, mining can change the pattern of surface water flow due to a lack of recharge from groundwater, the seepage from a stream bed, a decrease in the size of the catchment area or an increase in flow causing scouring or the inundation of points of limited capacity downstream. The quality of flow can be affected if there is poor containment of contaminants or ineffectual discharge.

Groundwater can be affected by the removal of topsoil, overburden and minerals and their replacement with contaminated or imported materials. This in turn may change quality, subsequently infiltrating water recharging the aquifer. The timing and relative rates of aquifer recharge and surface water flows are affected this way.

Discharges can cause physical and chemical contamination and the dewatering of workings will change the supply of water to abstractions, spring fed water courses and wetlands.

Technical advice on these matters is essential in the mineral planning process.

Proximity to communities

Hitherto, there has been a general view that planning control over mineral working has afforded greater protection to the natural environment than to communities. In any event, the relationship between a mine and it's host community now has a far greater profile than ever before. Indeed, there has in the past been a view shared by some operators that, provided they do not actually create a statutory nuisance, they are doing all that is expected of them.

However, I endorse the view that the quality of the residential environment would be much poorer if each new development was allowed to cause environmental deterioration to a point just short of statutory nuisance; desirable environmental quality demands much higher standards to be met.

In addition, it is important for decision makers to realise that the harmful effects of mineral working rarely, if ever, occur in isolation. There is a need to consider the cumulative impact of the effects and this may well be more than just an arithmetic addition.

7 THE NEED FOR MINERALS IN AN ERA OF SUSTAINABLE DEVELOPMENT

There are now many definitions of the phrase 'sustainable development'. It is, I firmly believe, a phrase of great relevance to the extraction of minerals, especially in environmentally fragile areas. The concept of 'living off the earth's income, rather than eroding it's capital' and 'handing down to successive generations not only material wealth, but also natural wealth' are concepts enshrined in mineral planning in Scotland.

As a rule, the environmental consequences of mineral extraction are difficult to reverse; the test for sustainability in relation to mineral extraction is whether the manmade wealth from mineral extraction justifies the consumption of finite resources and the environmental disruption it so causes.

The United Kingdom's sustainable development strategy, in relation to mineral extraction, seeks to conserve minerals as far as possible. Waste production must be minimised, there is now an emphasis on recycling to reduce primary demand, sensitive working practices are expected of developers and areas of critical landscape or nature quality are protected unless it is demonstrated that there are exceptional circumstances in the public interest for that development to proceed

Government policy nevertheless recognises that there should be an adequate and steady supply of minerals to meet the demands of the market place. As an example, Scotland consumes around 5.5 million tonnes of coal each year for power generation purposes. In a world of de-regulated markets, low priced, locally produced, quality coals contribute to the national energy supply, avoiding balance of payments deficits from imported overseas coals.

In it's commitment to putting sustainability at the heart of policy making, to the Government the matter of need is separated from the environmental considerations integral to determining a planning application.

In relation to energy minerals such as coal, the Government has not set limits or targets for any particular source or level of energy supply; nor to pre determine the levels of coal that must be produced from deep mined and opencast methods. Rather, in Scotland it is up to individual generating companies to determine the output for which they wish to aim, in the light of market conditions. It is a matter for the planning authorities to determine the acceptability of each proposal on it's own merits, having regard to environmental and planning policy issues.

Similarly with other minerals including aggregate, on-shore oil and gas, non construction aggregates such as limestone, silica sandstone and dimension stone and metalliferrous and specialised minerals, the determination of the best means of meeting the demands of the market is not an objective of the planning system: such decisions are to left to the commercial judgment of the minerals industry.

A sustainable minerals industry means therefore understanding the extent of mineral resources, not just in terms of quality and location, but also in relation to environmental constraints.

8 THE HARRIS SUPERQUARRY: A CASE STUDY IN SUSTAINABLE MINERALS PLANNING ?

Located just 200 km further south than Cape Farewell in Greenland, but some 4000 km almost due east, the remote Western Isles of Scotland have been the scene of Scotland's longest running minerals planning case.

Here, at Lingerbay on the southern tip of the Isle of Harris, Lefarge Redland Aggregates proposed to quarry some 550 million tonnes of igneous rock (anorthosite), at around 12-15 million tonnes per annum over a period of 60 years from a site that would extend to around 459 hectares.

The planning application was submitted in March 1991; the local Council was minded to grant planning permission in June 1993. The then Secretary of State called in the application and a public inquiry lasted between October 1994 and June 1995. In

April 1999, the Reporter who presided over the public inquiry recommended that planning permission should be granted; Scottish Ministers, in the newly devolved Scottish Government, refused planning permission for the development in November 2000.

Although Scottish Minsters have refused planning permission for the 'coastal superquarry', in late December 2000 an appeal was lodged by the developer in the Scottish courts on the basis that the Ministers' decision was unlawful. It may still be many months since the due legal process on Lingerbay is finally settled.

However, setting aside the current legal issues, Lingerbay is a prime example of the difficulties faced in dealing with major mineral working proposals in a very sensitive environment.

Lingerbay has a large consistent rock mass of suitable geology at a coastal location adjacent to deep water. The wedge shaped outcrop extends some 2 km by 5 km and comprises of some 95% Plagioclase Feldspar. Lingerbay is the only large outcrop of anorthosite in the UK and is very similar to anorthosite intrusions found in Scandinavia and eastern USA. Lingerbay aggregate would satisfy all tests to meet the requirements of road construction, sea defenses, high quality concrete and railway ballast.

The development of the quarry would include the building of a purpose built harbour capable of loading ships weighing up to 70,000 tonnes; at it's maximum production the quarry could employ around 100 people in an area characterised by high unemployment and out migration of the economically active workforce.

The highest tier of landscape designation in Scotland covers the 40 'national scenic areas' which are of 'unsurpassed attractiveness and must be preserved as part of our natural heritage'.

Lingerbay is situated in the second largest NSA which extends to 115,600ha and covers the islands of North Uist, South Lewis and Harris.

In environmental terms, objections to the quarry concerned a wide range of issues. The development would be in an environmentally sensitive and protected area which is required to be safeguarded; there would be permanent and severe damage to the landscape, an industrial development of an enormous scale would be introduced into an area of traditional industries such as farming, inshore fishing and tourism; protected bird species inhabit the site; marine life would be affected by surface run off and the 400 or so shipping movements each year would exacerbate marine pollution, especially from ballast water discharges. There was a risk of introducing contamination into pristine sea water which supported thriving local fisheries and shellfish poisoning was a particular concern. The effects of noise, dust and blasting on the amenity of local people was an issue, as were the social effects on the local Gaelic culture, particularly given the influx of workers that the site could require.

To those supporters of the development, the presence of a much needed and high quality mineral adjacent to sheltered deep water was a very important consideration. Similarly, there would only be minimal impact on marine and terrestrial wildlife and on tourism and local fisheries. A ballast water management plan would address many of the concerns about marine pollution; there would be benefit to local communities in the form of employment and the economic multiplier effect. The development would be in the national interest and the harm to the protected area would be offset by local and national economic benefits.

In refusing to grant planning permission, the Scottish Ministers disagreed with the approach adopted by the Reporter who presided over the lengthy public inquiry.

In particular, much of the need case advanced by the developer focused on the demand for aggregate for the congested south east of England. However, the Scottish Ministers were of the opinion that the planning system has no control over the eventual market to be served from mines and quarries and that production from Scottish 'superquarries' cannot be directed exclusively or even partly to English markets.

Moreover, the findings of the Inquiry confirmed that there was no evidence which could lead to the conclusion that refusal of the project would seriously prejudice the legitimate national interest in the continuing supply of aggregate sufficient to support national economic growth.

Landscape considerations prevailed. Indeed, despite recommending approval of the proposal, the Reporter at the Inquiry found that there would be a 'very significant' effect on the landscape; the quarry would be readily visible as a large and intrusive feature; it would completely change the landscape characteristics of the area by introducing a very significant industrial activity into the small scale pattern of a rocky indented coast and islets. In short, the quarry would create an area of massive disturbance and the acceptance of such an intrusive feature would undermine the successful operation of national scenic area policy.

In their decision, Scottish Ministers accepted that the development could contribute to economic development by direct and indirect job creation. However, the recommendation of the Inquiry seriously under-estimated the impact of the quarry on the environmentally sensitive protected area. To the Ministers, the objectives of designation and the overall integrity of the protected area would be compromised and the social and economic benefits did not clearly outweigh the significant adverse effects on the quality for which the area has been designated.

The scale and consequences of the Lingerbay case are immense. In the context of exceptionally complex planning procedures and environmental, economic and legal arguments there are clear implications for sustainable mineral planning in Scotland. Here a very clear message is being sent that, despite economic forces, the need to protect environmentally sensitive areas takes precedent.

9 ENVIRONMENTAL ASSURANCES IN MINERAL WORKING

When proposals for mineral working are deemed to be acceptable, and this does happen in by far the majority of cases, it is nevertheless imperative that the development proceeds in an environmentally acceptance manner.

Normally consent for mineral development is issued subject to a range of conditions. Such conditions cover everything from noise, dust and blasting control, landscaping, phasing, restoration and working methods. Separate consents address water and some air discharges. However, the proper control of mineral workings is essential, in the interests of the environment and host communities. Recent developments in this field point to improved practice and assist in the reduction of environmental costs and disbenefits.

Environmental management systems
Although a tool primarily for the mining and quarrying industry, the adoption of an environmental management system is a signal to regulatory bodies and the wider public that that a developer carries out his business to an accepted standard. As a response to increasing environmental awareness generally, a company that scrutinises it's own environmental performance and the potential impacts of it's actions is a positive step.

An environmental management system assists operational management to meet current and expected environmental obligations. It is a quality assurance system which can be used to measure company operations against environmental performance indicators.

Matters to be included in an EMS can include community consultation, emissions and performance monitoring, corporate environmental policy and regulatory and legal compliance verification.

ISO 14001 is the international standard for EMS and is, I believe, a vital tool for mineral working in the 21st century.

Environmental audits and monitoring
A regular environmental audit of a mining development assures the accuracy of environmental monitoring and demonstrates that the conditions attached to a development consent are being met. I would normally expect that a condition requiring at least an annual environmental audit was part of every planning permission. Normally the audit would be carried out by an independent assessor on behalf of the regulator but financed by the mining company. Matters to be assessed include the monitoring of noise, dust and blasting emissions, progressive restoration and the treatment of mine wastes. Complaints and breaches of control can be recorded and systems put in place to report and follow up difficulties.

ISO 14001 includes provision for environmental audits; the European Communities External Management and Audit Scheme (EMAS) is a voluntary system establishing a standard of environmental performance.

A more informal but nonetheless equally effective form of audit is to require the developer to fund the appointment of a 'compliance officer'. Here, an independent assessor will check the project on say, a monthly basis to ensure that the project is being carried out in accordance with approved plans, and that any necessary variation, and it's environmental consequences, are alerted to the regulatory bodies.

This task is assisted if the permission requires the submission of a 'mine progress plan' every 6 months or so, accompanied by a 'mine programme plan', setting out the ensuing phase of the development.

This system of checking allows deviations from originally prepared plans to be assessed properly within the context of the permission to work the site.

Performance and guarantee bonds
It is a sad reflection of earlier mining activity that there is frequently a legacy of dereliction and contamination left long after a mining project has been completed or otherwise abandoned. In this day and age this simply should not happen.

I believe it is essential that before a development is allowed to proceed, that the mining company and/or landowner has lodged with the regulatory body a 'performance bond' that can be used to restore a site and remedy any other environmental liability in the case of default by the developer.

The amount of the bond requires to be set by an individual arbiter and is usually lodged with the regulatory authority by an accredited bank or insurance company before the permission to work is issued.

Bonds can be flexible over the lifetime of a project and will increase or decrease as the amount of disturbed land or environmental liability changes. They should cover the full cost of restoring the site to the preferred state and need to cover provision for aftercare, professional fees and tendering procedures.

Where mining is located in a particularly sensitive area and despite the precautionary principle, there remains a residual risk to the environment, for example in terms of contamination of ground water, it seems reasonable that a mining company takes out environmental impairment insurance. Accordingly, if an adverse environmental event occurs after the cessation of a mining activity and perhaps long after a mining company has left the area, finance is available to address any problems without any expense falling on the public purse or, worse still, by the problem being ignored

Reviews of standards

By it's very nature, a mineral planning permission can last for a very long time. In the 15 or 20 year lifespan of such a consent, it is difficult to impose controls that will be as relevant in year 10 as in year 1. In Scotland, mineral planning permissions are subject to a statutory review every 15 years; I believe however that far more effective control is afforded if, say, every 3 years a company gives an undertaking to voluntarily review the standards included in a planning permission.

Community participation

A proposal that seeks to takes the local community with it from it's inception stands a greater chance of being accepted than one that is sprung upon an unsuspecting settlement without pre-application discussions. If a community can play a part in the design of the project and have it's say in environmental standards there will be a clear demonstration of a caring developer. Much is now written about community benefit and many council's and operators have in place a 'community trust' scheme whereby developers pay a local royalty on defined quantities of mineral extracted; this in turn is used for local environmental measures that can offset the presence of a mining company in a locality.

Properly constituted liaison groups, administered by a neutral local authority, can go some way in reconciling competing interests and are seen as a forum for discussing matters of local concern.

10 CONCLUSIONS: MINING INTO THE 21ST CENTURY

Environmental awareness is on the ascendancy, and so it should be. Working practices must be improved upon. Our landscapes remain scarred by the harmful effects of abandoned mines and quarries which are unsightly, can be dangerous and can give rise to harmful and often unregulated environmental effects long after mining has finished.

Sustainable development now demands that the best assets of our bio-sphere are passed down to future generations in no less a condition than that existing in our time.

Deciding on a mineral planning proposal is no easy task. I have been doing this professionally for over 15 years and it is getting tougher all the time as each party's expectations for their own agenda continue to increase. Achieving planning permission for even an average size project can be a costly business and could well cost well in excess of £100,000. And then there is no guarantee of success.

I believe that it is critical that those parts of our planet that are especially recognised for their natural heritage assets are given the proper protection that they deserve. Even where important minerals co-exist with treasured landscapes or important bio-diversity attributes, there must be a presumption against their extraction unless that deposit is otherwise so unique and that there is simply no alternative, in the national interest.

Particular care must be taken in assessing the environmental consequences of mining and it is important that full publicity and consultation is carried out and responses are taken into account.

Where development is to proceed, the role of the decision maker does not finish; auditing, monitoring and reviewing of a mining development are as integral to the scheme as the initial environmental assessment and industry must expect to pay for this regulation of it's actions.

The planning system must work hand in hand with the minerals industry; there is no place for conflict although it will continue to arise. Constructive dialogue between developers and host communities, with an emphasis on co-operation and an avoidance of an adversarial relationship must become the norm. For companies a trained and environmentally aware staff, with a flexible and understanding approach to environmental and community concerns will become increasingly important.

11 REFERENCES

Link Quarry Group 1996 *The Case Against the Harris Superquarry.* ISBN 0 9506157 4 9

Norman, C. 'Recent Developments in Planning Control over Surface Mining in Scotland'. *Proceedings of the Institute of Quarrying South Africa* Durban March 1999

Norman, C. 'Planning and Public Perception'. *Proceedings of the Institute of Quarrying* Bristol October 2000

Scottish Office Development Department 1994; *NPPG 4 'Land for Mineral Working'*

Scottish Office Development Department 1999; *NPPG 14 'Natural Heritage'*

Scottish Office Development Department 1999; *NPPG 16 'Opencast Coal and Related Minerals'*

Scottish Office Development Department 1996; Planning Advice Note 50, and annexes A-D, *'Controlling the Environmental Effects of Surface Mineral Working'*

Scottish Executive 2000 *'Report into the Lingerbay Public Inquiry'* vols. 1-4

Scottish Executive November 2000 press release *'Harris Superquarry Application Refused'*

Mining in the Arctic, Olsen, Lorentzen & Rendal (eds), © 2001 Swets & Zeitlinger, Lisse, ISBN 90 5809 177 5

MINEO. Use of hyperspectral data for monitoring pollution from the lead-zinc mine, Mestersvig, in Northeast Greenland

Peter Aastrup & Mikkel P. Tamstorf
National Environmental Research Institute, Dpt. of Arctic Environment. Frederiksborgvej 399. P.O. Box 358. Dk-4000 Roskilde. Denmark

Tapani Tukiainen
GEUS, Geological Survey of Denmark and Greenland, Thoravej 8. 2400 Copenhagen NV. Denmark

ABSTRACT: Presently seven European countries are studying the opportunities for using hyperspectral data to monitor pollution from mines in different climatic and vegetational regimes. The project is called MINEO and is supported by the EU. The Greenland study site is the former lead/zinc mine in Mestersvig, Northeast Greenland, which was operated 1956-1963. Pollution is well documented and levels of a number of metals in soil and vegetation will be studied and compared with hyperspectral data obtained during an airborne hyperspectral survey in august 2000. Long term effects of pollution on the composition of arctic vegetation will also be studied at the arctic site. It is expected that use of hyperspectral data will give valuable knowledge of the extent of pollution and will be an important tool for planning of monitoring programs.

1 INTRODUCTION

Monitoring of pollution from mines is important to assess potential health risks for humans, wildlife ingested by humans, and plants. Monitoring involves chemical analyses of samples of soil and organisms close to the mine site as well as far away from the mine site. Monitoring sites are usually selected to be representative for detection of elevated levels of substances in any direction and distance from the mine. Wind directions, topography and atmospheric conditions highly influence the distribution of pollution. Therefore it is necessary to establish a relatively large number of sampling stations, and to analyze a large number of samples.

Remote sensing techniques offer the possibility to get an overview of large areas at a relatively low cost. Analysis of hyperspectral data has given promising results for detection of surfaces with specific characteristics such as elevated levels of e.g. heavy metals. It is foreseen that environmental monitoring will be improved and costs minimized by using hyperspectral data.

The MINEO project was initiated in January 2000 and involves six mine sites in Portugal, Germany, Austria, Great Britain, Finland and Greenland. This paper presents the Greenland test site.

The MINEO project is partly financed by the EU (Information Societies Technology (IST) Programme) and partly by a number of governmental institutes and private companies in Europe.

2 BACKGROUND

It is expected that hyperspectral data will be available from the OrbView-4, PRISM, NEMO and HRST sensors within a few years. In order to be prepared to use this kind of data an airborne scanning of hyperspectral data from a variety of geological, vegetational and climatic regimes in Europe and Greenland was performed in summer 2000. This will be followed up by field studies in summer 2001.

3 STUDY AREA

The Greenland site is the former lead-zinc mine in Mestersvig in Northeast Greenland at 72.18° N and 24.17° W. Mestersvig is located in the high arctic with sparse vegetation. Pollution from the mine, which was closed in 1963, is documented (Agger *et al.* 1991; Asmund *et al.* 1997) and logistically the area is relatively easy to reach from Europe by small planes suitable for scanning flights.

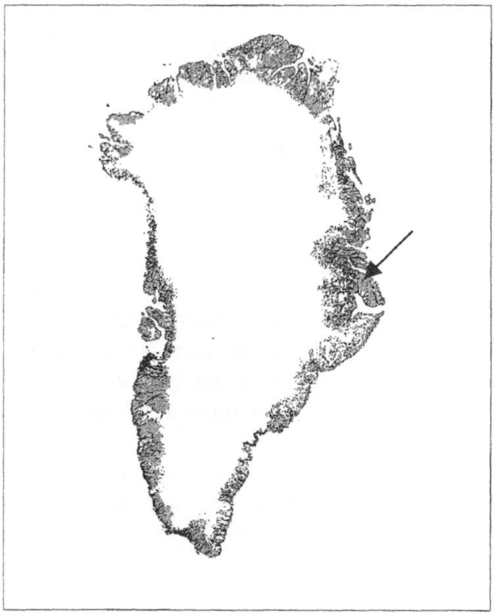

Figure 1. Greenland. Mestersvig shown with an arrow.

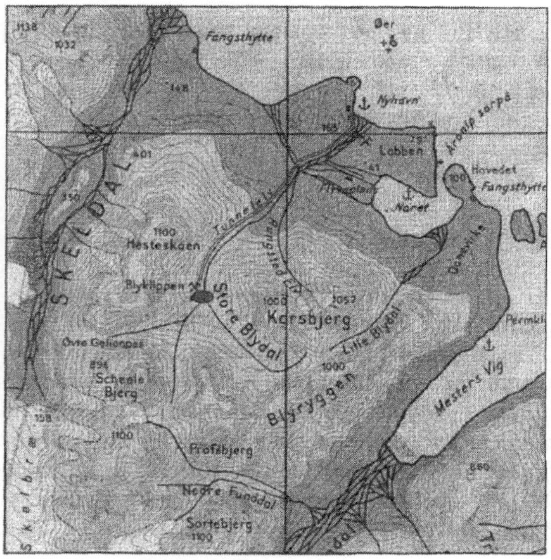

Figure 2. Study area. The mine site is indicated with a dot.

4 OBJECTIVES

The MINEO objectives are for studies in Mestersvig are:

1) to analyze hyperspectral data from the Mestersvig area.
2) to document levels of lead, zinc, and other metals in soil and plants from areas indicated by hyperspectral data to have elevated levels of metals.
3) to sample spectra from surface types, which judged from airborne data seem to be polluted.
4) to analyse and compare vegetation in polluted areas and vegetation in unpolluted areas.
5) to collect data for a spectral library of surface types.

5 METHODS

Hyperspectral data

Hyperspectral data from the test sites, including Mestersvig, were collected during the summer of 2000. The data were collected with a HyMap scanner that was flown on a Dornier 228 at an approx. altitude of 6000 f.a.g. (Figure 3). The HyMap scanner, built by integrated Spectronics Inc. of Sydney Australia, has four spectrometers in the region .45 to 2.45 nanometers excluding the two major atmospheric water absorption windows. The bandwidths are not constant, but vary between 15 and 18 nm.

Figure 3: The Dornier 228 at Konstapel Pynt in Greenland with the HyMap scanner and Zeiss camera installed (Photo: B.U.Hansen).

The scanner is spectrally and radiometrically calibrated in the laboratory before and after the campaign or during a campaign if rough handling damages its alignment. The scanner has also an on board bright source calibration system that is used to monitor the stability of the signal. The noise level of the data measured outside the aircraft with a sun angle of 30° and a 50% reflectance standard was more than 500/1 except near the major atmospheric water absorption bands. Figure 4 shows flight routes in the Mestersvig area.

Air photos

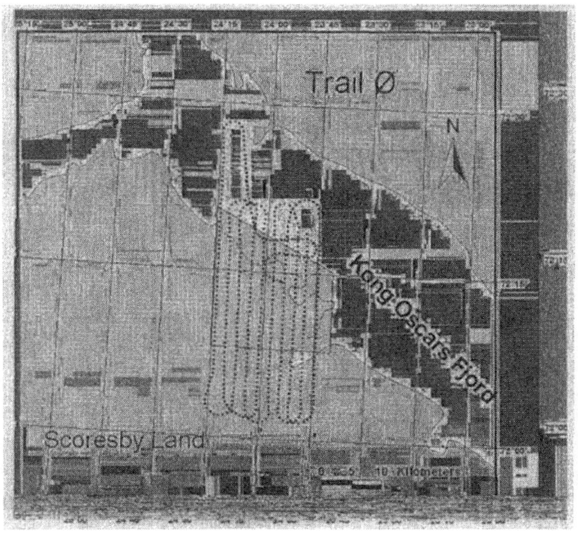

Figure 4. Flight routes in the Mestersvig area

Digital terrain information was obtained from air photos with a Zeiss-RMK A 15/23 aerial camera which were acquired simultaneously with the hyperspectral data. The air photos enable the production of a digital terrain model with a spatial resolution better than 1 meter and will be used for the geometric correction of the hyperspectral data.

Ground truth
Fieldwork to obtain ground truth for the hyperspectral data will be carried out during the summer of 2001. This fieldwork will be done in order to facilitate a link between the hyperspectral data and in situ levels of pollution and plant cover.

The fieldwork will be concentrated at 30 sites where plants and sediment will be analyzed for levels of lead, zinc, barium and several other metals which can be directly related to the mining process. The location of the sites will be based on an analysis of the distribution of the mentioned elements from the hyperspectral data. Sites will be placed in areas that indicates high levels of pollution as well as in areas that show no or little indications of pollution. In this way we should be able to define the minimum level of pollution that it is possible to detect be hyperspectral data.

Spectral signature of the plant cover and sediment will be measured and used to develop a reference spectral library. The library can later be used as ground truth for the hyperspectral data analysis and for modeling of the distribution of the plants and pollution. Figure 5 shows spectral signatures for some minerals and green grass is shown as reference for the vegetation signature. Minerals normally differ significantly in the mid-infrared spectra between 2.2 and 2.5 µm were the different minerals have distinct peaks or troughs at individual wavelengths. The most distinct pattern of vegetation is around 0.75 µm where chlorophyll and internal cell structure cause the jump in reflection.

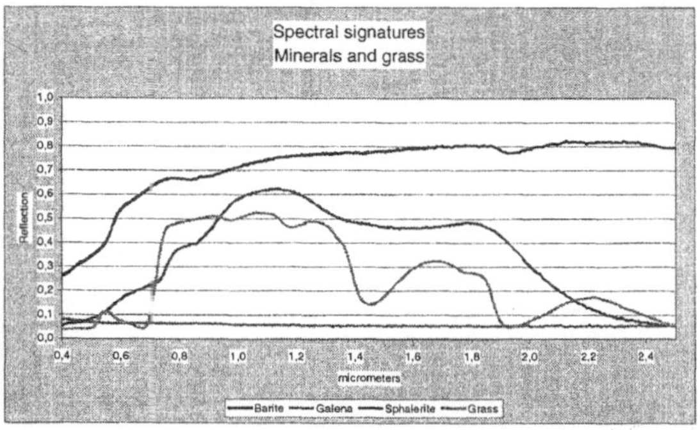

Figure 5. Spectral signatures of minerals and grass

Processing
The hyperspectral data will be orthorectified based on the digital terrain models and corrected for atmosphere.

The analysis of the distribution of pollution is based on the theory of linear unmixing. The spectral signatures of all the different surfaces within the smallest unit in the image, the pixel, combine the spectral signature of this pixel. If the surface of this pixel is a combination of half barite and half sphalerite the resulting spectral signature will be a combination of their signatures, each weighted by a half. If all the occuring clean spectral signatures (end members) inside a pixel is known, it is possible to estimate the abundance of each endmember by linear unmixing.

6 PERSPECTIVES

It is expected that the use of hyperspectral data will greatly increase the possibilities to delineate polluted areas. Figure 6 shows an example of lead pollution in sand from the shore in the Mestersvig area. It is based on samples collected along the shore, and it can be seen that the level of lead in sand is increasing with distance to the north and to the south from the source "Kaj". "Kaj" is Danish for "Warf" and is situated near the anchor inside the small inlet named "Naret" on the map of the study area (Figure 2). In this case the traditional method would demand more samples at greater distances from the "Kaj". We hope that hyperspectral data will help us to delineate the pollution by revealing elevated levels. This will have at least two important perspectives: Monitoring costs will decrease and delineation of pollution will be much better defined. In addition to this, we hope to assess long term pollution effects on vegetation types.

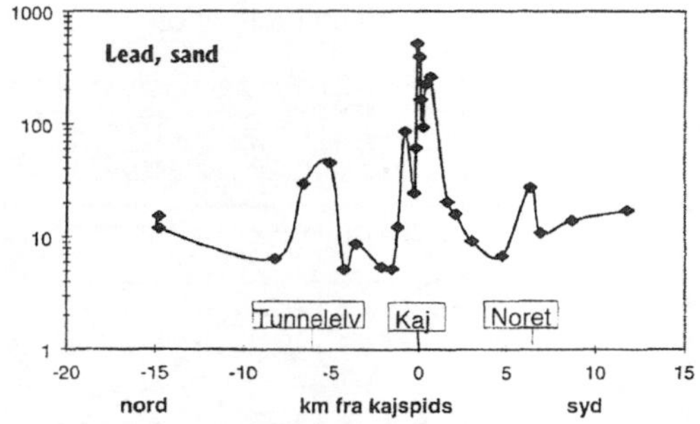

Figure 6. Level of lead in sand along the shore in the Mestersvig area. The x-axis shows distances to the north and south from the source named "Kaj" in km (Asmund *et al.* 1997).

7 REFERENCES

Agger, C. T., G. Asmund, R. Dietz & P. Johansen 1991. *Miljøundersøgelser ved Mestersvig 1991.* Grønlands Miljøundersøgelser 1991, 23 pp (in danish).

Asmund, G., F. Riget & P. Johansen 1997. *Miljøundersøgelser ved Mestersvig 1996.* Faglig Rapport fra DMU, nr 202, 31 pp (in danish).

Mining in the Arctic, Olsen, Lorentzen & Rendal (eds), © 2001 Swets & Zeitlinger, Lisse, ISBN 90 5809 177 5

Pollution from Mining in Greenland – A Review

Poul Johansen & Gert Asmund
National Environmental Research Institute, Department of Arctic Environment, Frederiksborgvei 399, 4000 Roskilde, Denmark

ABSTRACT: In Greenland mining started as early as the 1850s. Environmental protection in connection with these operations was not an issue dealt with until the early 1970s, when the largest mine in Greenland was to start operating. Historically three major mines have been in production, a cryolite mine in South Greenland, a lead-zinc mine in East Greenland, and a lead-zinc mine in West Greenland. These mines all have created heavy metal pollution. The most severe of these have been pollution with lead and zinc, when tailings, waste rock or concentrate have got into contact with sea water. In particular the simultaneous action of waves, tidal water and air seems to enhance dissolution of lead and zinc. In contrast, waste rock and tailings deposited on deep water do not cause lead to be released, whereas this is the case for zinc. In one case dust from an ore crusher and from a conveyor belt carrying metal concentrates was also an important source, causing metals to be dispersed widely through the atmosphere. Results from environmental monitoring studies and monitoring in water, sediments and biota at the three mines are

1 INTRODUCTION

In Greenland mining started as early as the 1850s and has taken place for copper, graphite, coal, cryolite, lead, zinc and silver. Environmental protection in connection with these operations was not an issue dealt with until the early 1970s, when the largest mine in Greenland was to start operating.

Since then a legal framework has been developed for environmental protection in connection with exploration and exploitation of minerals in Greenland. Monitoring and assessment of environmental impacts has been part of this process.

In this paper we discuss the environmental impacts of mining operations in Greenland. The paper is mainly based on the publications shown in the reference list, in particular a publication by Johansen and Asmund (1999).

Historically three major mines have been in production: A cryolite mine in South Greenland (section 2), a lead-zinc mine in East Greenland (section 3) and a lead-zinc mine in West Greenland (section 4). The location of these is shown in Figure 1.1. We focus on heavy metal pollution, which has been the far most important environmental problem of the mines.

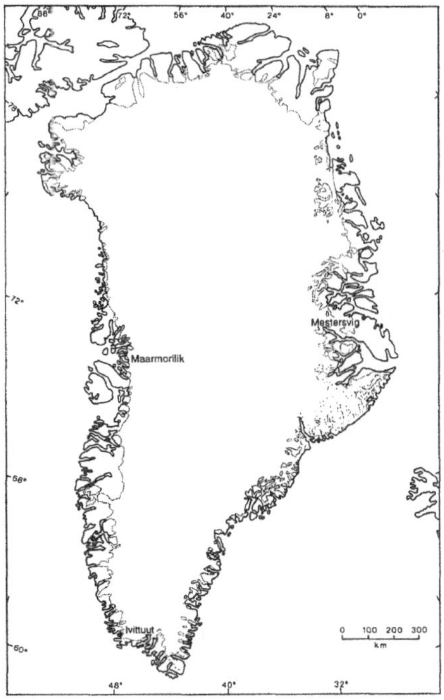

Figure 1. Location of the three major mines in Greenland (Maarmorilik, Ivittuut, and Mestersvig).

2 THE CRYOLITE MINE AT IVITTUUT

This mine is located at Ivittuut in Arsuk Fjord in South Greenland, see Figure 1.1. A small Greenland community, Arsuk with c. 300 inhabitants is located 15 km to the west of Ivittuut, and Greenlanders from this community use Arsuk Fjord for fishing and hunting. Kangililinnguit (or Grønnedal) 5 km to the north east of Ivittuut is a Danish naval station with c. 200 inhabitants, mainly Danes who use the area for recreation.

2.1 *The mining operation*

Ore was mined from 1854 to 1987 by the Danish company Kryolitselskabet Øresund A/S. It was an open pit operation. 3.3 million tonnes of cryolite ore was shipped out during the lifetime of the mine, most of this in the last 50 years, when about 50,000 tonnes were shipped out annually. The cryolite was located in one orebody, c. 115 x 50 metres in plan and c. 70 metres deep. The main part of the orebody consisted of siderite-cryolite rock, i.e. cryolite with about 20% siderite and 1-2% quartz and sulphides. The formation was located very close to and partly in the fjord, and a dam was built towards the fjord to prevent seawater from filling the pit.

At Ivittuut the ore was handled only mechanically and was shipped out after crushing and sorting. Flotation of the ore was not carried out in Greenland but at a plant in Copenhagen, Denmark. During the first many years of mining only almost totally "clean" cryolite with a very low content of other minerals was shipped out. Therefore large quantities of "unclean" cryolite were produced and it was used as landfill in quay and road building. Later the methods to extract cryolite from the ore were improved, so that ore with a cryolite content down to 40% could be used. This meant that it became profitable to use the old landfill areas containing cryolite, and in the later part of the lifetime of the mine extensive digging, sorting and shipment of cryolite from landfill areas took place. Low-grade ore originally left in the pit was also produced during the last few years of the mine's life.

2.2 *Pollution sources*

The first environmental studies at Ivittuut were carried out in 1982. They showed that lead and zinc had polluted seaweed and blue mussels in the intertidal zone of a large part of the fjord (see section 2.3). This was unexpected, and further studies were initiated to identify pollution sources. The far most important was shown to be the land-fill areas at the coastline, which released dissolved lead and zinc under influence of the tidal water, whereas other possible pollution sources, dust dispersal and water release from the pit, were not important.

The chemical composition of the landfill areas has been studied as part of an exploration project to mine them. On average they contain 12% cryolite, 2200 mg/kg lead and 2040 mg/kg zinc. In addition to lead and zinc, copper and cadmium also occur, but in lower concentrations, and seem not to have affected biota in the fjord.

A study was initiated to quantify the transport of dissolved lead from the landfill areas at the coast by sampling and analysing water from five different levels in the landfill and at different times over the tidal period. This was repeated several times over a whole year. The general finding was that seawater with a lead concentration of about 0.1 μg/l entered the landfill areas with the tide. In the landfill lead was released to the entering tidal water resulting in lead concentrations between about 20 and 5000 μg/l dissolved in the seawater that returned to the fjord at falling tide. It was estimated that the amount of lead released from the landfill areas was between 1 and 3 kg per day corresponding to between 400 and 1000 kg per year. A similar but not as extensive study was conducted for zinc, and it was concluded that the release of zinc was in the same order of magnitude as for lead.

2.3 *Environmental impact*

Environmental studies have shown that pollution with lead and zinc is found in fjord sediments and in the intertidal zone. Brown seaweed's and blue mussels have been used as indicators of lead and zinc pollution of the fjord, and monitoring studies since 1982 have shown a decreasing trend over the monitoring period is observed. A possible explanation of this trend could be that lead and zinc gradually are washed out by the tidal action and removed from the landfill areas.

The highest metal levels are found at the pollution source in Ivittuut with decreasing values away from the source. In blue mussels lead is elevated above background on a coastline of more that 50 km, whereas the affected area for seaweed is small (Fig. 5.1).

3 THE LEAD-ZINC MINE AT MESTERSVIG

This mine is located at Kong Oscars Fjord in East Greenland, see Figure 1.1. It is far from human settlements, the closest one being the small Greenland town Illoqqortormiut (or Scoresbysund) about 200 km to the southeast with c. 500 inhabitants. They live mainly from hunting marine and terrestrial animals. However they only rarely visit the Kong Oscars Fjord area at the mine.

3.1 *The mining operation*

The mine and a concentrator was underground, located c. 10 km inland from Kong Oscars Fjord, see Figure 3.1. It was operated by the Danish company Nordisk Mineselskab from 1956 to 1963 when the orebody was emptied. Then 554,000 tonnes of ore had been mined and 58,000 tonnes of lead concentrate and 75,000 tonnes of zinc concentrate produced. A road was constructed between the site and the fjord, where concentrate was shipped out. Tailings from the concentrator were discharged at the mine on a nearby mountain slope, from where most has slid into the river Tunnelelv flowing to Kong Oscars Fjord.

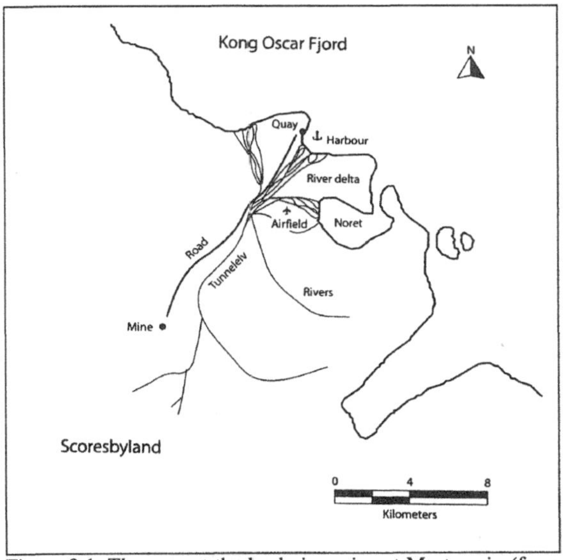

Figure 3.1. The area at the lead-zinc mine at Mestersvig (from Johansen and Asmund 1999).

3.2 *Pollution sources*

The operation seemingly was carried out without any thoughts about environmental impacts. The first environmental studies were carried out in 1979. They showed that tailings were an important pollution source. It was estimated that only 15% or 66,000 tonnes of tailings remained in the original tailings disposal area, and that most of the tailings had been washed into the river, the river deltas and the fjord. The tailings remaining had high concentrations of lead (0.66%) and zinc (2.13%).

The environmental studies also identified the loss of concentrate in connection with loading operations in the local harbour at the fjord as an important pollution source. During the operating period, concentrate was transported in bags from the mine to the harbour. Here they were loaded on barges, which were hauled to a ship anchored a few hundred metres off the coast. The bags were then opened and emptied into the ship. It is known that significant amounts of concentrate were spilled in the fjord during loading, including whole 4 tonnes bags and whole barge loads of about 30 tonnes of concentrate.

3.3 *Environmental impact*

As mentioned the first environmental studies were conducted in 1979. They showed as would be expected high levels of lead and zinc in the sediments of Tunnelelv and in the river deltas at Kong Oscars Fjord (up to 13% Zn and 0.7% Pb in the finest sediment fraction (< 50 μm)). Also high concentrations of lead and zinc were found in the water of Tunnelelv downstream the mine (zinc values up to 150 μg/l dissolved and 397 μg/l particulate, lead values up to 12 μg/l dissolved and 25 μg/l particulate).

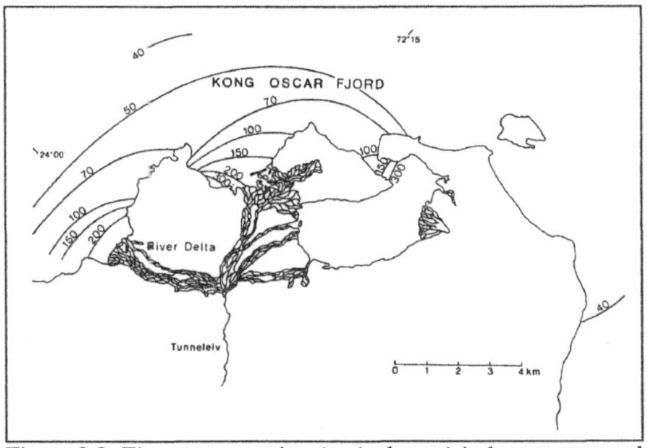

Figure 3.2. Zinc concentration (μg/g dry wt) in brown seaweed at Mestersvig in 1979. Lines combine points at the coast with equal concentration (from Johansen and Asmund 1999).

The most significant impact was found in brown seaweed from the intertidal zone. Elevated lead and zinc levels were found on a significant part of the south coast of Kong Oscars Fjord, see Figure 3.2. The figure shows that the heavy metal pollution of the fjord originates from three main areas: The northern and the southern delta of Tunnelelv and the harbour area in-between reflecting the transport of tailings in Tunnelelv and the spill of concentrate at the harbour. In seaweed a significant part of the coastline, more than 30 km, has elevated metal levels (Fig. 5.1).

The pollution has been monitored since 1979 and both lead and zinc concentrations have decreased over the monitoring period as a whole. However, lead levels increased significantly in 1991. This was caused by erosion of a severely lead polluted area in the harbour, resulting in exposure and release of old concentrate spillage.

Overall levels are expected to continue to decline, as the contaminated material becomes more and more dispersed and gets mixed up with or covered by non-polluted sediments. However, Kong Oscars Fjord will be affected by heavy metal pollution for many years. Lead and zinc concentrations in the marine environment may be expected to fluctuate with high levels following coastal erosion of major contaminated land areas, followed by rapidly decreasing levels in periods with no erosion.

4 THE LEAD-ZINC MINE AT MAARMORILIK

The mine was located at Maarmorilik in the inner region of the large Uummannaq Fjord system in Northwest Greenland, see Figure 1.1. A small Greenland settlement, Ukkussissat, with c. 200 inhabitants is found about 25 km from the mine. People in this region of Greenland mainly live from fishing and from hunting marine mammals and birds, but while the mine operated also several local people worked here.

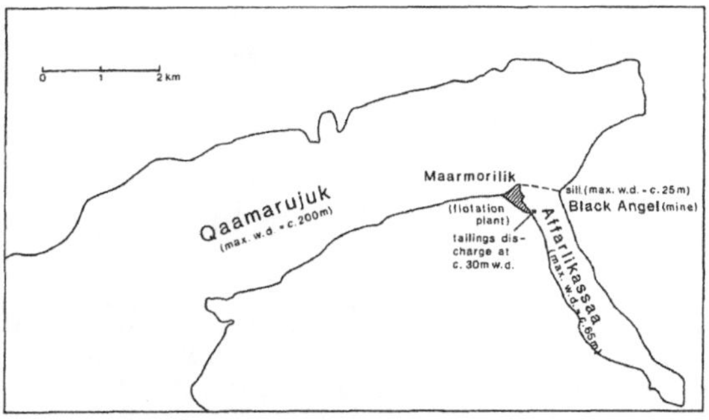

Figure 4.1. The Maarmorilik area (from Johansen and Asmund 1999).

4.1 *The mining operation*

The Danish company Greenex operated the mine called the Black Angel from 1973 to 1990. It has been the largest mining operation in Greenland. About 11 million tonnes of ore with c. 12% zinc and 4% lead was mined and milled at Maarmorilik.

The general outline of the operation is shown in Figure 4.1. The main ore bodies were found in the Black Angel Mountain 600 metres above sea level. Crushed ore was brought in cable cars across a fjord to a flotation plant in Maarmorilik, where on average 135,000 tonnes of zinc concentrate and 35,000 tonnes of lead concentrate were produced annually and shipped out in the open water season (June to December). Tailings from the plant were discharged into the small adjacent fjord Affarlikassaa.

Before the mine started operating in 1973, the environmental impact was assessed and environmental baseline studies were conducted. Shortly after the mine began producing, studies showed that the environmental impact was much more severe than anticipated, and since then environmental monitoring and regulations became part of an ongoing process to mitigate the impact.

4.2 *Pollution sources*

The main pollutants were identified to be lead and zinc. Trace amounts of mercury, cadmium and copper also occurred, but not in concentrations so high as to cause important elevations of levels in marine biota. Several sources of heavy metal pollution were identified: Tailings, waste rock, ore crushing and transport of concentrates were the most significant. Polluted water from the mine also occurred, but only for a short period.

In terms of quantities of pollutants released to the environment, tailings were by far the most important source. Tailings were discharged at 25-metre depth into the fjord Affarlikassaa. Here solids settled, as a sill at the mouth of the fjord prevented tailings solids from entering the adjacent larger fjord Qaamarujuk (see Figure 4.1). However, soon after mine start in 1973 we found highly elevated concentrations of dissolved lead and zinc in the waters of Affarlikassaa. When the fjord water mixed during winter, dissolved metals were transported from the lower to the upper water column and over the sill into the surface waters of Qaamarujuk. It is likely that lead and zinc carbonate in the tailings were the source of metal dissolution, although most of the metals in the tailings occurred as sulphides. It was estimated that between 10 and 30 tonnes of lead and between 30 and 55 tonnes of zinc dissolved per year while the mine operated. After settling of the tailings on the fjord bottom only very small amounts (if any) of lead were released to the overlaying water, whereas zinc was released in significant amounts, about 10 tonnes per year.

A second important pollution source is waste rock dumps. During the first ten years of operation, disposal of waste rock outside the underground mine workings was not regulated, and 2-3 million tonnes of waste rock with 0.1-0.8% lead and 0.3-2.3% zinc were left in four dumps on the slopes of steep mountains in the area. Metals are released to the sea mainly in the form of sulphide particles in rain and melt water from glaciers. An average annual input of 8 tonnes of lead has been estimated from this source. Most of this, however, settles close to a river outlet in the fjord.

One of the waste dumps was located directly at the Qaamarujuk Fjord, partly in and below the coastline. This caused a high mobilisation of lead and zinc from the waste, which was in direct contact with seawater and was exposed to wave and tidal action. The highest lead and zinc levels in seaweed and blue mussels were found at this dump. As part of a plan to mitigate metal pollution in the area after mine closure, this dump material was removed to the extent possible, and mainly deposited on top of the tailings at the bottom of Affarlikassaa.

A third significant metal pollution source was dust. An annual input to the environment of 2 tonnes of lead and 5 tonnes of zinc from all dust sources has been estimated during the last years of the mine operation. The main dust source probably was the ore crusher, and because it was located c. 600 metres above sea level, dust and metals from it could be dispersed over long distances. Another important dust source was the conveyor belts transporting concentrates from the concentrate storage building to the bulk carriers.

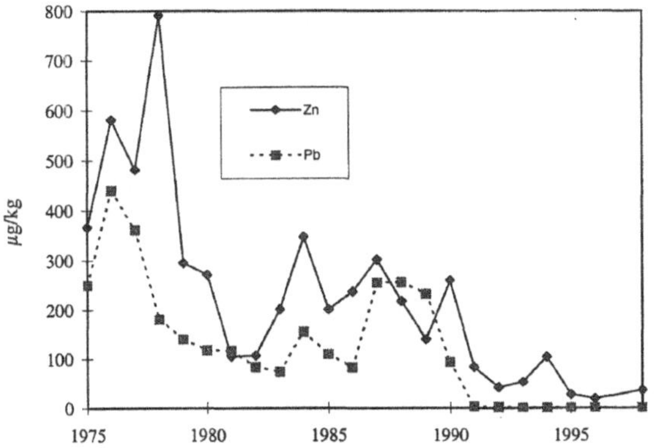

Figure 4.2. Zinc and lead dissolved in the bottom water of Affarlikassaa.

4.3 *Environmental impact*

As mentioned the environmental impact turned out to be much more severe than anticipated shortly after the start of the mining operation. For many years the impact has been monitored closely.

Seawater from the fjords at Maarmorilik has been analysed for lead and zinc on a regular basis. Figure 4.2 illustrates the lead and zinc concentration in the bottom water of the Affarlikassaa Fjord into which tailings were discharged.

It is seen that the highest metal levels were found in the beginning of the period. In the late 1970s both the lead and the zinc concentration decreased in the fjord as a result of increased flotation efficiency and other mitigative actions. From then the zinc concentration remained at the same level until mine closure after which it has declined significantly, but deposited tailings and waste rock on the bottom of Affarlikassaa still release zinc, as was expected. After an initial significant decline, the lead concentration remained at a lower level for about 10 years, but in 1987 it increased because a new orebody which was more difficult to flotate was brought into production. After mine closure in 1990 the lead pollution of seawater has changed drastically, as may be seen in Figure 4.2, and only very small amounts of lead are released from settled tailings and waste rock on the bottom of Affarlikassaa. While mining took place, lead and zinc levels were also elevated in Qaamarujuk Fjord, particularly in winter when mixing of the water masses in Affarlikassaa caused a transport of dissolved lead and zinc into the surface waters of Qaamarujuk.

Brown seaweed was analysed for lead and zinc as biological indicators of the pollution in space and time. The main results for seaweed are shown in Figure 5.1. It is seen that seaweed's from a large part of the coast of the fjords have elevated zinc and lead concentrations and that levels have decreased over the monitoring period, particularly for lead. Zinc levels have remained rather high after mine closure compared to lead.

In blue mussels lead but not zinc rose to very high levels at the mine after the start of the operation. Lead concentrations above 3000 μg/g (on dry weight basis) were observed at the most affected sites. This is more than 3000 times higher than the concentration found in unaffected areas. Lead levels in blue mussels were elevated in all three fjords at the mine. In this area - up to about 30 km from the mine - it has for many years been recommended not to collect and eat blue mussels because of high lead concentrations.

A number of marine fish species and a prawn species have been monitored regularly. Lead levels in the meat have remained low and mostly not elevated, whereas in liver and bone tissue elevated lead levels have been found in many cases. The lead levels in fish and prawns have generally decreased since the mid 1980s and particularly after mine closure. Lead concentrations in seals, birds and in fish meat have all been low and below limits for safe consumption in human food.

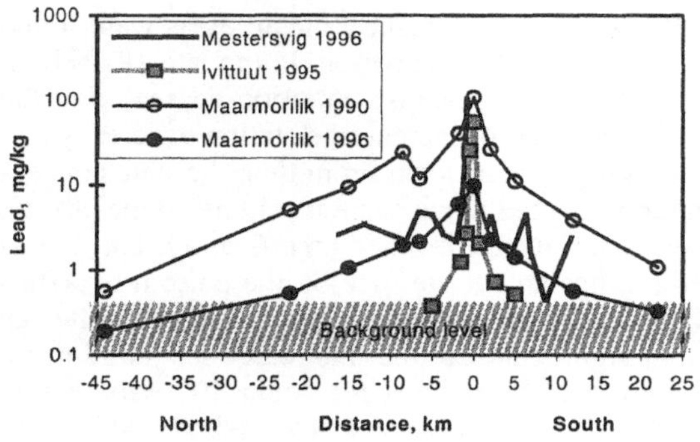

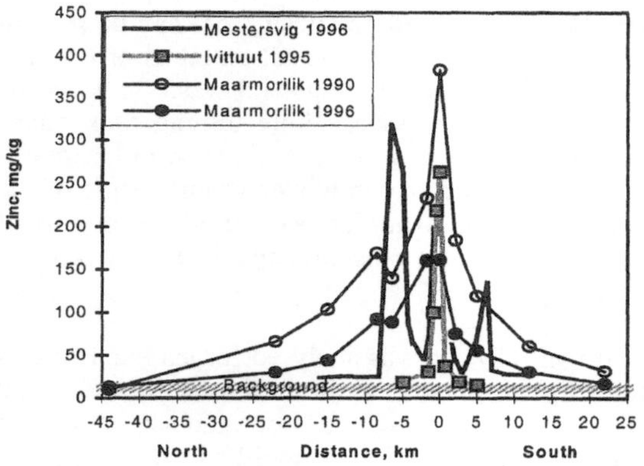

Figure 5.1 Concentrations of lead and zinc in growth tips of seaweed (*Fucus spp.*) around the three major Greenlandic mines. Also shown are background concentrations found elsewhere in Greenland (from Riget et al. 2000).

5 CONCLUSIONS

The three major Greenland mines all have created heavy metal pollution. The most severe of these appears to be pollution with lead and zinc, when tailings, waste rock or concentrate have got into contact with sea water. In particular the simultaneous action of waves, tidal water and air seems to enhance dissolution of lead and zinc. In contrast, waste rock and tailings deposited on deep water do not cause lead to be released, whereas this is the case for zinc. Figure 5.1 compares the pollution by the three mines by showing the lead and zinc levels found in seaweed.

This knowledge and other experience from environmental monitoring at mine sites in Greenland is very important and will be used when dealing with future mine projects. Environmental protection is an important issue when new projects are evaluated and is part of a legal framework for approving and regulating these.

At new projects the operators are responsible to assess associated environmental impacts and to design their operations in a way so that impacts are avoided or limited to a degree acceptable to the regulating agency. Environmental baseline and monitoring studies are important tools in this process. At the end of operations sites must be restored to the extent possible and with no significant risks of pollution from deposited waste. Plans for restoration when closing an operation must be prepared already at the application stage of a project.

REFERENCES

Agger, C.T., Asmund, G., Dietz, R., Johansen, P. (1991) Miljøundersøgelser ved Mestersvig 1991. Teknisk Rapport. Grønlands Miljøundersøgelser, 23 pp. *(Environmental studies at Mestersvig 1991. Technical Report. Greenland Environmental Research Institute, 23 pp)*

Asmund, G. (1980) Water movements traced by metals dissolved from mine tailings deposited in a fjord in northwest Greenland. Pp. In Freeland, H.J., Framer, D.M. and Levings, C.D. (Eds.) Fjord Oceanography. Plenum Press, New York and London, 347-353

Asmund, G. (1992a) Pollution from the marine tailings disposal at the lead-zinc mine at Maarmorilik, West Greenland. In: Singhal RK, Mehrotra AK, Fytas K, Collins J-L (Eds) Environmental Issues and Waste Management in Energy and Minerals Production. Balkema, Rotterdam, pp. 587-594

Asmund, G. (1992b) Lead and zinc pollution from marine dumping of waste rock from lead/zinc mining. In: Bandopadhyay S, Nelson MG (Eds) Mining in the Arctic. Balkema, Rotterdam, pp. 105-112

Asmund, G., Broman, P.G., Lindgren, G. (1994) Managing the environment at the Black Angel Mine, Greenland. Int J of Surface Mining, Reclamation and Environment 8: 37-40

Asmund, G., Christophersen, J.B., Steensboe, J. (1990) Rehabilitation and demolition after the closure of the zinc and lead mine "Black Angel" at Maarmorilik, Greenland. Proc. Polartech 90, Copenhagen, pp. 774-759

Asmund, G., Hansen, M.M., Johansen, P. (1988) Environmental impact of mineralized waste rock at Maarmorilik and Ivittuut, Greenland. In: Proc. from Int. Conf. on Control of Environmental Problems from Metal Mining. Røros. State Pollution Control Authority, Oslo Norway

Asmund, G., Hansen, M.M., Johansen, P. (1988) Environmental impact of marine tailings disposal at the lead-zinc mine at Maarmorilik, West Greenland. Proc. from Int. Conf. on Control of Environmental Problems from Metal Mining, Røros. State Pollution Control Authority, Oslo Norway

Asmund, G., Johansen, P. (1999) Short and long term environmental effects of marine tailings and waste rock disposal from a lead/zinc mine in Greenland. In R. Fernandez Rubio (Ed) Proceedings: International Mine Water Association "Mine, Water & Environment" International congress September 13-17 Sevilla Spain, pp 177-181

Asmund, G., Johansen, P., Fallis, B.W. (1991) Disposal of Mine Wastes Containing Pb and Zn Near the Ocean: An Assessment of Associated Environmental Implications in the Arctic. Chemistry and Ecology 5: 1-15

Asmund, G., Johansen, P., Riget, F. (1997) Miljøundersøgelser ved Mestersvig 1996.Faglig rapport fra DMU, nr. 202, Miljø- og Energiministeriet, Danmarks Miljøundersøgelser, 31 pp *(Environmental studies at Mestersvig 1996. Technical Report no. 202. Ministry of Environment and Energy. National Environmental Research Institute, 31 pp)*

Glahder, C., Asmund, G., Stijl, F.vd (1996) Natural zinc elevations in Arctic water bodies. In: Ciccu R (Ed) Environmental Issues and Waste Management in Energy and Mineral Production. SWEMP '96. DIGITA University of Cagliari pp 811-817

Hansen, M.M., Asmund, G. (1986) Miljøundersøgelser i Kong Oscars Fjord 1985. Grønlands Fiskeri- og Miljøundersøgelser, Grønlands Geologiske Undersøgelse, 41 pp *(Environmental studies in Kong Oscars Fjord 1985. Greenland Fisheries and Environmental Research Institute, Greenland Geological Survey 41 pp)*

Johansen, P., Asmund, G. (1999) Pollution from Mining in Greenland: Monitoring and Mitigation of Environmental Impacts. In Azcue J.M. (Ed.) Environmental Impacts of Mining Activities, Springer-Verlag, Chapter 15, pp 245-262.

Johansen, P., Asmund, G., Riget F (1995) Miljøundersøgelser ved Ivittuut 1982-1992. Rapport Serie nr. 7, Grønlands Miljøundersøgelser, 60 pp *(Environmental studies at Ivittuut 1982-1992. Report Series No. 7. Greenland Environmental Research Institute, 60 pp)*

Johansen, P., Hansen, M.M., Asmund, G. (1985) Heavy metal pollution from mining in Greenland. In: Fernández-Rubio R (Ed) Mine Water Congress, Proc Sec Int Congress, Granada Spain, pp 685-693

Johansen, P., Hansen, M.M., Asmund, G., Nielsen, P.B. (1991) Marine Organisms as Indicators of Heavy Metal Pollution - Experience from 16 Years of Monitoring at a Lead Zinc Mine in Greenland. Chemistry and Ecology 5: 35-55

Johansen, P., Riget, F., Asmund, G. (1997) Miljøundersøgelser ved Maarmorilik 1996. Faglig rapport fra DMU, nr. 193, Miljø- og Energiministeriet, Danmarks Miljøundersøgelser. 97 pp *(Environmental studies at Maarmorilik 1996. Technical Report no. 193. Ministry of Environment and Energy. National Environmental Research Institute, 97 pp)*

Loring, D. H., Asmund, G. (1987) Heavy Metal Contamination of a Greenland Fjord System by Mine Wastes. Environ.Geol.Water Sci. 14(1): 61-71.

Møller, J..S. (1984) Hydrodynamics of an Arctic fiord. Inst. of Hydrodynamics and Hydraulic Engineering, Technical University of Denmark, Series Paper No. 35. Lyngby, Denmark.

Riget, F., Johansen, P., Asmund, G. (1997) Uptake and release of lead and zinc in blue mussels. Experience from transplantation experiments in Greenland. Mar Poll Bull 34: 805-815.

Riget, F., Johansen, P., Dahlgaard, H., Mosbech, A., Dietz, R., Asmund, G. (2000) The Seas around Greenland. In Sheppard C (Ed.) Seas at The Millennium: An Environmental Evaluation. Pergamon, Elsevier Science, pp. 5-16.

Mining in the Arctic, Olsen, Lorentzen & Rendal (eds), © 2001 Swets & Zeitlinger, Lisse, ISBN 90 5809 177 5

Control strategies for acid mine drainage in Arctic regions

B. Elberling
Institute of Geography, University of Copenhagen, Copenhagen, Denmark

ABSTRACT: Mining in permafrost-affected areas raises unique issues that need to be addressed to minimize the environmental impact of mining. The exploitation and disposal of sulfidic tailings can lead to acid mine drainage (AMD) including acidification and leaching of heavy metals. Permafrost features that influence the deposition of sulfidic waste include processes within the active layer that produce pathways for reactant and contaminant transport, barriers such as snow and ice covers, and low temperatures that slow most chemical and biological processes. Primary issues for disposal of sulfidic mine tailings are summarized and strategies for controlling the oxidation of sulfide minerals at sub-zero temperatures are discussed. Results from a case study in Canada illustrate the possibilities of keeping tailings permanently water-saturated or frozen, thereby controlling AMD. Future research should focus on AMD generation rates at sub-zero temperatures and solute migration in frozen media.

1 INTRODUCTION

The main environmental concern facing the mining industry today is the occurrence of dissolved heavy metal contaminants and acidic effluents caused by sulfide oxidation (Paktunc 1999), generally referred to as acid mine drainage (AMD). There are numerous abandoned, operating and proposed base metal mines in permafrost-affected regions, and the disposal of sulfidic mine waste and AMD are problems known at several mine sites in the Arctic. Mine tailings which result from the processing of ore are usually transported to nearby disposal areas and acidic drainage will in most cases cause metals to spread to watershed stream systems and the marine environment.

Pyrite (FeS_2) is by far the most common sulfide mineral giving rise to AMD. The oxidation of pyrite is a complex biogeochemical process involving several redox reactions, hydrolysis, complex ion formation, solubility control and microbial catalysis. Although several aspects of this process are not fully understood, the following equation of reaction is commonly given to describe the complete oxidation of the iron and sulfur moieties of pyrite:

$$FeS_{2(s)} + 3.75O_{2(g)} + 0.5H_2O \rightarrow 2SO_4^{2-} + Fe^{3+} + H^+ \quad (1)$$

In the reaction, ferric iron and sulfuric acid are formed and heat is released. Recent reviews on the process and controls are available (e.g., Evangelou 1995). It is well known that aerobic sulfide-oxidizing bacteria can catalyse the oxidation of sulfide minerals and numerous studies describe the catalysing effects of chemolithotrophic bacteria such as *Thiobacillus ferrooxidans* (Evangelou 1995, Nordstrom & Southam 1997). These bacteria are in particular known to be able to catalyse the oxidation of sulfide minerals (Eq. 1) under highly acidic conditions, thereby increasing the overall oxidation rate by orders of magnitude as compared to abiotic reactions (Nordstrom & Southam 1997). As well as the degree and type of microbial activity, the following factors are generally thought to influence AMD generation: position of the sulfide body within the hydrological regime, types and abundances of sulfides and neutralizing gangue minerals, grain size distribution (specific surface area), climate, internal tailing temperatures and oxygen availability.

The first part of this review (sections 2 and 3) will discuss the influence of bacteria, temperature and oxygen on AMD generation as these factors play a unique role in permafrost regions. In addition, freezing-point depression, unfrozen water content and solute redistribution will be considered, since these factors influence the environmental impact of AMD. The last part of the review (sections 4 and 5) will focus on literature reporting observed AMD in permafrost-affected areas and discuss the current knowledge on strategies for controlling AMD.

2 BIOTIC SULFIDE OXIDATION AT SUB-ZERO TEMPERATURES

In the permafrost region, temperatures are low for most of the year and temperature fluctuations can be extreme over short periods of time. Temperature is an important control on the physical and chemical reactions that are the basis for all biogeochemical processes. Therefore, subsurface temperature is considered one of the most important parameters regulating the activity of bacteria within the soil environment. Bacteria that have adapted to live in cold conditions and to survive periods of extreme low temperatures have been known for more than 100 years (Forster 1887). Since then, several surveys have documented the presence of viable bacteria in permafrost environments (e.g., Morita 1975). Cold-adapted microbes have been classified as either psychrophiles (cold-loving, optimal temperatures for growth below 15°C, active to below zero) or psychrotrophs (cold-tolerant, optimal temperatures for growth above 15°C, active to below zero). One of the key conditions for the presence of viable microbes at freezing temperatures is the presence of liquid water. In frozen soil, thin intergranular water films allow chemical processes to continue even in winter and may represent survival niches for microbial life. The winter survival strategies for bacteria in the Arctic are not well understood, but bacteria are able to survive extremely low temperatures and viable micro-organisms occur in permafrost to depths of at least 18 m and in material older than 70,000 years (Boyd & Boyd 1964, Kjøller & Ødum 1971).

2.1 *The presence of sulfide-oxidizing bacteria in permafrost regions*

The number of studies addressing the presence and potential activity of psychrophilic or psychrotrophic sulfide oxidizing bacteria in High Arctic environments is limited. Kalin (1987) reported the presence of psychrophilic thiobacilli at the Nanisivik Mine, Baffin Island, Arctic Canada. Psychrotrophic iron- and sulfide oxidizing bacteria, e.g., *Thiobacillus ferrooxidans*, have been detected in alkaline tailings at Nanisivik Mine (Elberling et al. 2000). In the latter study, micro-calorimetric measurements revealed that as much as 35% of the overall oxidation of tailings (measured as heat production) was due to the presence of the bacteria. Langdahl (1999) isolated a psychrotrophic acidophilic sulfide-oxidizing microbial population (dominated by *T. ferrooxidans*) from naturally exposed sulfide ore material in North Greenland (83°N).

The presence of the well known sulfide-oxidizing bacteria *T. ferrooxidans* in High Arctic environments indicates that these bacteria populations are likely to be found at any location in the permafrost zone where reactive sulfide minerals are or have been present.

2.2 *Temperature-dependent biotic and abiotic sulphide oxidation*

Most biogeochemical reaction rates decrease with decreasing temperatures and are described well by the Arrhenius equation:

$$\ln (k_1/k_2) = E_a (T_1-T_2) / RT_1T_2 \qquad (2)$$

where k_1 and k_2 are reaction rates at temperatures T_1 and T_2, E_a is the activation energy of the reaction and R is the gas constant. Values of E_a for pyrite oxidation reported in the literature range from 39 to 88 kJ/mol under a variety of experimental conditions (Nicholson et al. 1988). Although the effects of temperature on sulfide oxidation rates can be complex, particularly under field conditions, several studies confirm that the simple relationship given by the Arrhenius equation is a valid approximation of chemical and of microbial reaction rates (Dawson & Morin 1996). However, the Arrhenius equation considers only a single activation energy that is not temperature-dependent, whereas complex biological systems are presumed to have several temperature-dependent forms.

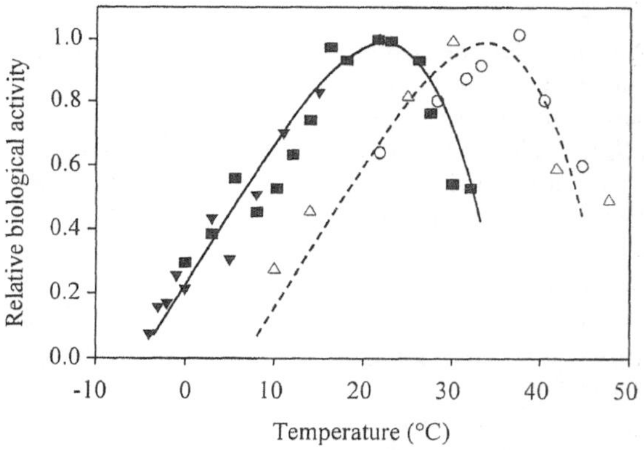

Figure 1. Relative changes in biological oxidation rates of pyrite collected from temperate latitudes (open marks) and arctic latitudes (filled marks). Changes in oxidation are given relative to the maximum oxidation observed. Open marks represent data from Jaynes et al. (1984), filled squares are from Langdahl & Ingvorsen (1997), and filled triangles are from Elberling (2000).

Figure 1 shows the temperature-dependent biotic activity of sulfide-oxidizing bacteria collected from arctic and temperate regions. The dashed line represents an equation fit (Jaynes et al. 1984) to relative activity measurements (X_t) of bacteria collected at temperate latitudes and considered valid for the temperature range 4° to 55°C (T):

$$X_t = -1.2\times10^{-5}T^3 - 4.4\times10^{-4}T^2 + 0.066T - 0.25 \quad (3)$$

The solid line represents a fit to bacteria collected in the Arctic. An overall difference in temperature optimum of 7-8°C is observed and indicates an important biological adaption to cold climate.

It is evident that metabolic processes are active at temperatures as low as -10°C (Gilichinsky et al. 1992). However, limited information is available on the growth of specific sulfide-oxidizing bacteria at sub-zero temperatures. Langdahl & Ingvorsen (1997) studied the production rate of dissolved Fe^{2+} and Fe^{3+} of an enriched *T. ferrooxidans* culture from naturally exposed sulfide ore material in North Greenland. After laboratory investigations it was concluded that the overall rate of microbial activity at 0°C constituted as much as 30% of the maximum activity at 21°C.

Oxygen uptake rates as a result of sulfide oxidation have been observed at temperatures as low as -4°C. These rates are based on the difference in oxygen uptake in ambient and sterilized tailing samples from Nanisivik Mine (Elberling 2000). Results from that study reveal a uniform decline in biotic and abiotic oxygen uptake rates in pyritic tailings with decreasing temperatures (Fig. 2). The temperature-dependent oxygen uptake rate was consistent with the Arrhenius relation in the temperature range above 0°C.

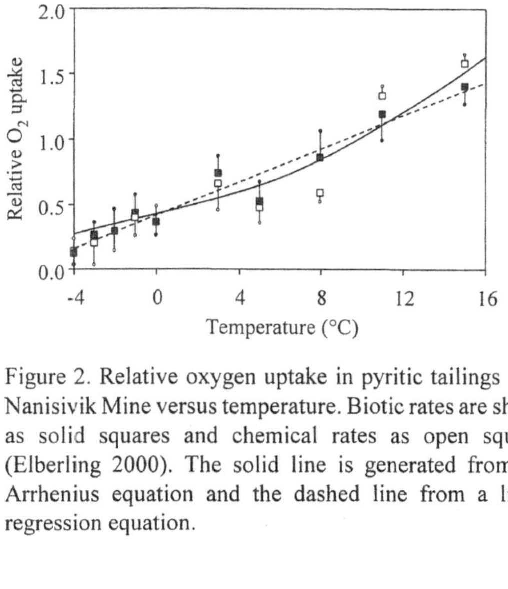

Figure 2. Relative oxygen uptake in pyritic tailings from Nanisivik Mine versus temperature. Biotic rates are shown as solid squares and chemical rates as open squares (Elberling 2000). The solid line is generated from the Arrhenius equation and the dashed line from a linear regression equation.

Below 0°C an increasing deviation is observed and the linear correlation provides the best fit for the entire temperature interval studied. Significant variations between the dependency of biological and chemical oxidation on temperature were not observed, indicating that the influence of temperature on the overall AMD is not sensitive to the degree of biological oxidation. Meldrum et al. (1999) studied the oxygen uptake in tailings from Rankin Inlet in Arctic Canada, in freezer experiments at temperatures as low as -10°C. The authors conclude that oxygen consumption and heat generation due to oxidation are negligible at -10°C. The few studies mentioned above indicate that bacteria in natural systems may to a large extent be adapted to activity at sub-zero temperatures. Although the observed bacteria activity reviewed above indicates a significant reduction at 0°C compared to maximum activity, the results suggest that biological sulfide oxidation may take place at temperatures as low as -4°C, and that the activity becomes negligible as temperatures approach -10°C. Although the kinetics of sulfide oxidation show significant potential AMD at sub-zero temperatures, sulfide oxidation will only occur if oxygen and liquid water are present. The next section deals therefore with the potential transport and redistribution of oxygen and water in frozen tailings.

3 TRANSPORT IN FROZEN GROUND

Frozen sediments consist of several phases including solid particles, ice, unfrozen water, and gas. Within the frozen sediment, ions may be transported as a result of a concentration gradient (diffusion) or by a pressure gradient (advection). Extreme differences in the diffusivity within the various phases play a major role in the control of the overall AMD generation and spreading of contaminants. Ion diffusion in pure monocrystalline ice is considered to be insignificant as ion diffusion coefficients are about five orders of magnitude lower than in water (Barnall & Slotfeld-Ellington 1983). For components present in the gas phase, e.g., oxygen, the diffusion coefficients are about four orders of magnitude higher in the gas phase than in the water phase. Consequently, atmospheric oxygen will reach the depth where sulfide oxidation takes place primarily as a result of diffusion through the partially gas-filled pores (e.g., Pantelis & Ritchie 1992, Elberling et al. 1993, 1994). Only for waste material stored under near-saturated or fully-saturated conditions the lower oxygen diffusion in the water phase becomes important.

3.1 *Transport of oxygen*

As pyrite oxidizes in tailings, atmospheric oxygen is consumed and the change in concentration with depth results in a concentration gradient, which is the driving force for the diffusive transport of oxygen. In tailing systems, one-dimensional molecular diffusion of oxygen gas can be described by Fick's First Law:

$$F_D = -D_e\, dC/dz \quad (4)$$

where F_D is the flux of O_2, D_e is the effective diffusion coefficient, and dC/dz is the vertical concentration gradient for oxygen. D_e is a function of the diffusion coefficient for oxygen in air and a factor that depends on the percentage of gas-filled pores (relative to the volumetric moisture content) and the tortuosity (Troeh et al. 1982). Several models are available for estimating D_e but few are based on tailing material. Assuming that water-filled and air-filled pathways represent parallel resistance to diffusion in tailing material, Elberling & Nicholson (1996) showed that the effective diffusion coefficient (D_e) in tailings can be accurately approximated as the weighted sum of the diffusion coefficients in air and water (see dashed line in Fig. 3).

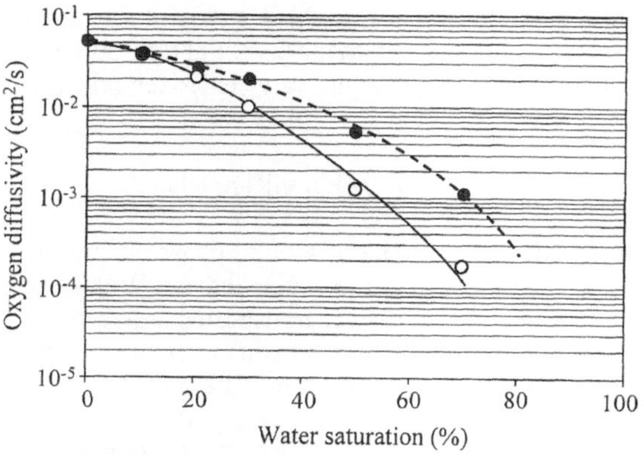

Figure 3. The effective diffusion coefficient in non-frozen tailings (filled circles and dashed line) and frozen tailings (open circles and solid line) at water saturations up to 70%. The dashed line is an empirical equation for estimating D_e based on saturation values (Elberling & Nicholson 1996) while the solid line is a free-drawn line representing the observed effect of freezing for comparison (Elberling 1998).

In frozen media, diffusion estimates require oxygen diffusion coefficients for super-cooled water as well as estimates on the overall change in gas pore volume fraction and connectivity during freezing. Such data are currently not available from the literature, and diffusion coefficients for oxygen gas in unfrozen and frozen tailings with variable water saturation are only known from experimental designs. Diffusion measurements on unfrozen and frozen tailings from Nanisivik Mine have shown that freezing has no significant effect on D_e in tailings with a water saturation less than 30% (Elberling, 1998). Above 30% saturation, freezing lowers D_e, reaching a reduction factor of about 10 at 70% water saturation. It was concluded that freezing well-drained tailings deposited on land is not expected to decrease the overall availability of oxygen in tailings. The study points out that changes in the degree of saturation may be as important as freezing in permafrost areas and that a combination of saturation and freezing seems ideal for controlling oxygen availability in tailings deposited in the Arctic.

3.2 *Unfrozen water*

The amount of liquid water present in "partly" frozen soil and waste materials is critical for controlling microbial activity and oxygen diffusion. It also affects water and solute movement in frozen soils, which in turns influences soil strength, frost heaving and contaminant transport.

Solutes lower the freezing point of water during freezing and the relationship is given by the Van't Hoff equation. The freezing-point depression in dilute solutions can be calculated by using the following simplified equation (Marion 1995):

$$\Delta T \approx -1.86\, \upsilon\, m_b \qquad (5)$$

where ΔT is the freezing-point depression, υ is the number of aqueous species resulting from the dissolution of the solutes, and m_b is molarity of the solute. For pore water extracted from approximately 7-year-old pyritic tailings in Nanisivik (unpublished data) concentrations of magnesium and sulfate alone result in a freezing-point depression of more than 1°C compared to the corresponding freezing-point of pure water. However, deviations from Equation 5 are apparent for high concentrations, and particularly for electrolytes that dissociate into ions of higher valance.

A nearly pure ice phase will form when temperatures drop below the freezing-point, concentrating solutes in the remaining unfrozen solution and thereby further decreasing the freezing point. If equilibrium is maintained between the solution and ice phases during freezing the concentrations of various ions in solution will follow the ice-solution equilibrium line, increasing in concentration as the temperature decreases towards the eutectic composition. In a pure NaCl-H_2O system with an initial NaCl molality greater than 5.17 mol/kg, the solution will not freeze until it reaches the eutectic temperature of -21.2°C. Below this temperature the solution will solidify as a mixture of ice and salt (Marion 1995).

The chemical equilibrium and the unfrozen water content during freezing can be modelled by FREZCHEM (Mironenko et al. 1997) as this model includes the coefficients for the Pitzer equations as well as solute redistribution during freezing. However, under natural conditions, and particularly for tailing water, site-specific characteristics will influence the ion exclusion and thereby the unfrozen water content. These characteristics include texture, specific surface area, charged surfaces, freezing rate and the initial distribution of salts (Marion 1995), and are not included in the model. Within these limitations, FREZCHEM has been used to simulate the observed unfrozen water content measured in columns filled with mine tailings from Nanisivik (Elberling 2000). Unfrozen water content was measured by time domain reflectometry (TDR) and the tailings were samples from the upper reactive zone and partly oxidized zone (20 cm) as well as the lower reduced zone near the permafrost (80 cm). The chemical composition of the pore water from the same site and depths was used as input to FREZCHEM to predict the amount of unfrozen water content versus temperature (Fig. 4).

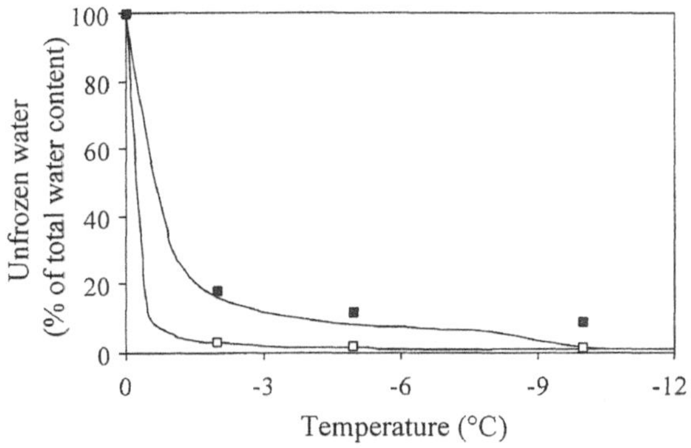

Figure 4. Unfrozen water content (% of total water content) in mine tailings collected at 20 cm within the oxidized zone (filled squares) and at 80 cm in the reduced zone (open squares) versus temperature. Observations are compared to FREZCHEM simulations (Elberling 2000).

The significant amount of unfrozen water, in particular in the reaction zone, suggests that liquid water is available for biological activity and contaminant transport well below 0°C.

3.3 *Water movement in frozen soil*

A potential environmental effect of the presence of unfrozen water is the diffusion of pollutants in the water phase and water movement within tailing profiles. The water movement is caused by the liquid films on particles which provide a route for water and solutes in frozen tailings (Godwaldt et al. 1999). In general, water in frozen sediments moves from warm to cold regions, from regions with low concentration to regions with higher concentrations and along water tension gradients (Marion 1995). As tailings in permafrost areas freeze from the top downward and from the permafrost table upward, the thermal gradients will induce a water flow towards the freezing fronts. The comparative importance of these two freezing fronts is determined during the freezing process by environmental factors such as water content, temperature and snow cover.

As solutes are largely excluded in the freezing process, maximum solute concentrations are generally found close to the freezing front. The solute concentration in front of the freezing front may rise to 80 times the original concentration if no solutes are included in the frozen zone, and thus cause the freezing front to leap over solute pockets (Hallet 1978). These solute pockets may ultimately freeze, resulting in alternating bands of high and low ion concentrations in frozen tailings. This ion exclusion in freezing sediment has received much attention and is shown to be controlled by temperature gradient (freezing rate), pore water composition, water content, mixing and sediment type (e.g., Konrad & McCammon 1990). The extent of ion exclusion increases with decreasing freezing rate (Marion 1995). For freezing rates smaller than 0.1°C/day, corresponding to most field freezing conditions, more than 90% of the solutes have been observed to be rejected from the pore ice formed (Konrad & McCammon 1990).

The ion exclusion may result in an increased environmental impact from AMD during springtime when thawing starts from above and causes highly variable ion concentrations in the meltwater. Ion pulses (meltwater with very high ion concentrations) are known from most arctic areas (Melak & Sickman 1995, Rasch et al. 2000). Such pulses will have a maximum impact on living organisms sensitive to toxic levels of for example, heavy metals as well as on the overall flux of contaminants from tailing areas.

Water flow in frozen soils containing pore ice can be described by Darcy's Law in which the gradient in liquid pressure is the driving force and the hydraulic conductivity the transport characteristic. Although this Darcy flow is coupled to water flow induced by gradients in electrical potential and solute concentrations, values of apparent hydraulic conductivity for frozen soil have been derived (summarized by Kay & Perfect 1988). Generally, the apparent hydraulic conductivity decreases by orders of magnitudes as a result of freezing. However, this conclusion is based solely on laboratory observations of porous media with low or moderate concentrations of ions and has not been verified for partly weathered tailings with extreme concentrations of weathering products. The extent to which a general understanding of transport properties in frozen ground can be used to predict the potential movement of contaminants in frozen tailings is therefore unclear.

3.4 *Soil strength*

Frost heaving occurs as a result of the formation of segregated ice, e.g., lenses and needles, as well as the volumetric expansion of water during freezing. Models are available to simulate and predict the mechanisms responsible for frost heaving (e.g., Chamberlain 1981, Black & Hardenberg 1991). Marion (1995) summarized the most important factors for the degree of frost heaving, including soil texture, pore size, freezing rate, temperature gradients, moisture and overburden stress. Frost heaving in tailings is considered a disadvantage as it produces cracks and preferential pathways that allow the infiltration of oxygen gas and oxygenated water. However, frost heaving is seldom observed in tailing areas, probably due to high ion concentrations that tend to reduce frost heaving. As pointed out by Sheeran & Yong (1975), this is to be expected because high ion concentrations increase the unfrozen water content and reduce the water available for forming segregated ice in tailings. Frost heaving may be important in cover materials provided as protection on top of tailings, and grain size and porosity should be considered in order to minimize frost heaving.

4 AMD IN PERMAFROST REGIONS

Temperature fluctuations in continuous permafrost regions cause the surface layer to thaw annually. The thaw depth, defined as the active layer, varies typically between <0.5 m under thick organics to several metres under mineral soil. Pyrite oxidation is known to occur naturally in the active layer in permafrost areas where mining activity has never taken place. Cameron (1977) studied the oxidation of a massive sulphide body named the Agricola Lake prospect located northeast of Yellowknife, Canada. Apparently, intensive sulfide oxidation was occurring in the active layer both near the surface and at depth, and the lack of carbonate minerals in the area had resulted in natural acidic drainage (pH ~3-4) and metal leaching. Cameron (1979) observed the same phenomena in a massive sulfide deposit on Melville Peninsula in Northwest Territories, Canada. However, the presence of calcareous outcrops neutralized the acidic drainage and surface water. Active oxidation in permafrost areas in Canada has been confirmed by more recent observations in Yukon, Canada (Kwong 1995).

Another example of naturally exposed sulfide body is at Citronen Fjord in North Greenland where the mean air temperature varies between -15 and -20°C and temperatures above zero are only recorded during a very short summer period. During this period, the *in-situ* oxygen uptake was measured directly in well-drained and ochre-coloured sediment below the exposed sulfide outcrops in order to estimate the sulfide oxidation rate (Elberling & Langdahl 1998). Oxygen uptake rates in the order of 0.2-0.4 moles/m^2 per day were observed during September 1996. As organic carbon was absent, the amount of oxygen consumed was assumed to represent the overall rate of sulfide oxidation. The high oxygen uptake rates observed were consistent with high concentrations of dissolved zinc and sulfate in pore water and surface drainage from the area.

The examples above illustrate that high sulfide oxidation rates occur naturally in continuous permafrost areas, at least during the short summer. Thus, weathering of naturally exposed sulfidic material may be important in relation to evaluating background concentrations of metals prior to mining activity.

4.1 *Oxidation of mine tailings deposited in continuous permafrost areas*

Various waste materials are produced during the mining process. Mine waste consists of large chunks of rocks or overburden which are transported by trucks or dragline to waste dumps. The ore material is transported to the mill where processing consists of grinding followed by the extraction of the metals such as Zn, Pb, and Cu. Mine tailings, the waste product of metal extraction, consist of fine-grained sulfidic ore material with a reduced content of heavy metals. Mine tailings and waste rocks have very different hydrogeological, geochemical, and geotechnical characteristics and must therefore be considered separately in relation to AMD. The following section only evaluates the disposal of mine tailings; a more thorough discussion of the disposal of waste rocks in permafrost areas is provided by Dawson & Morin (1996).

Mine tailings consist of ore material finer than 0.2 mm and comprises a high proportion of sulfide grains with a large, fresh surface area. The results of ongoing investigations of oxygen uptake and heavy metal release in pyritic mine tailings deposited on land near Nanisivik Mine on Baffin Island (NWT) in northern Canada (73°N) are used in the following to illustrate the controlling parameters for AMD generation in well-drained tailings deposited in the continuous permafrost region (Elberling 2000).

In Nanisivik, the annual mean temperature (1994-1998) is -15.6°C and the annual precipitation is approximately 125 mm. Since 1976 mine tailings have been deposited under water in a tailings impoundment. As the available subaqueous storage capacity was exhausted in 1991 disposal of tailings began on land. A test plot within this area was selected and *in-situ* oxygen uptake rates were estimated by changes in oxygen concentrations in closed gas chambers placed on top of the tailings (Elberling & Nicholson 1996). Measurements were made periodically throughout the year and were accompanied by climatic measurements. Figure 5 shows that by the end of the summer the active layer extended to a depth of approximately 2 m and that this zone freeze during autumn.

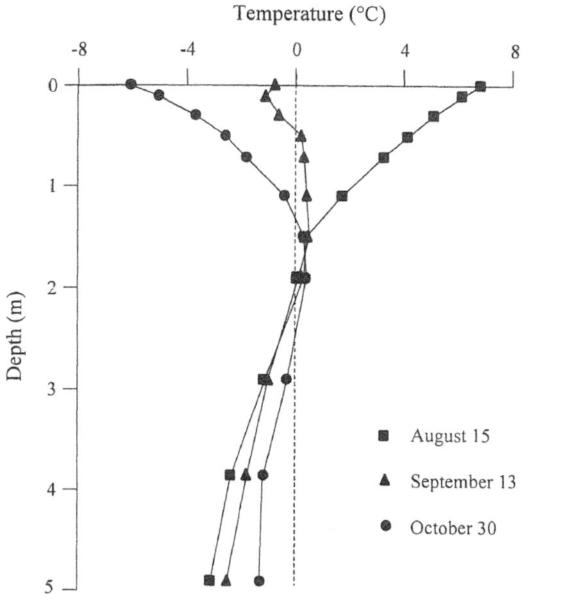

Figure 5. Soil temperature profile of an uncovered site at the tailings impoundment in Nanisivik. The tailings froze during a period of three months in 1997 (Elberling 1998).

However, the freezing of the active layer varies significantly between years depending on water content and snow accumulation that insulates the ground from low air temperatures.

During summer months, oxygen uptake rates in 6-year-old well-drained tailings were 0.2 to 0.5 moles O_2/m^2 per day (Elberling 2000), similar to rates observed in temperate climates (Elberling et al. 1996). During the following 2-months period, the air temperature dropped below 0°C, resulting in significant effects on soil temperatures and the observed oxygen uptake (Fig. 6). Soil temperatures were used to model temporal variation in the reaction constant k (Eq. 2), and subsequently the expected variation in oxygen uptake over time, assuming that the diffusion coefficient within the reaction zone remained constant (Fig. 6). The model failed to predict the oxygen uptake during the first days following a major precipitation event on Julian day 216 (11 mm over less than 10 hours). However, a site covered with plates to prevent rain water infiltration but allowing free oxygen diffusion made it possible to differentiate between the effects of temperature and increasing water content on the overall oxygen uptake. Figure 6 shows that the model can predict the effect of temperature but not the effect of decreasing oxygen diffusion. The observations reveal the importance of soil temperature and water content on AMD.

Measurements during winter revealed that temperatures in the upper part of the snow-covered tailings were in the range of -1 to -3°C and the oxygen consumption was about 0.12 ± 0.02 moles of O_2/m^2 per day or approximately 20-25% of the rates observed during summer. During early spring, temperatures the upper part of tailings were below -10°C and oxygen uptake was not detectable. The first oxygen uptake during spring was observed as the soil temperature rose to -4°C. These detailed measurements of oxygen uptake throughout the year indicate that AMD generation is highly sensitive to changes in temperature and that the effect of maintaining tailings below 0°C is predictable (Fig. 2).

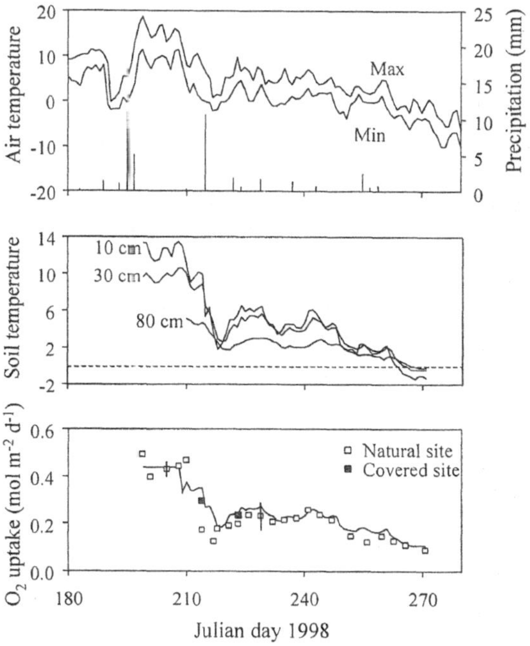

Figure 6. Air and soil temperatures (curves), precipitation (bars), and *in-situ* oxygen consumption rates at a natural site (open squares) and a covered site (solid squares) in well-drained tailings measured in 1998. Bars through squares represent the SD of replicates (n=6). The simulated oxygen consumption is shown as a solid line (Elberling 2000).

5 AMD CONTROL STRATEGIES

AMD control strategies are based on control of AMD generation, control of contaminant migration, or collection and treatment of contaminated effluent. Although collection and treatment of contaminated effluent is a proven and demonstrated control strategy, maintaining such a facility at remote northern minesites, following closure, is costly and logistically difficult (Dawson & Morin 1996). AMD generation and contaminant migration will be discussed in the following in relation to different control strategies, as these strategies aim to limit both AMD generation and contaminant migration.

5.1 *Frozen covers*

Permafrost regions offer a unique possibility for controlling AMD that is not naturally available in warm climates: the freezing of mine tailings by covering the tailings with an inert material and thereby forcing the frost table to move up and maintain the tailings at sub-zero temperatures all year round. The application of frozen covers has been tested in few places in continuous permafrost areas. At Lupin Mine, located 400 km NE of Yellowknife, Canada (65°N), sulfidic tailings (containing pyrrhotite, FeS) have been covered with 60 cm of gravel sand. Temperatures below the cover were above 0°C for more than 4 months each year and tailing temperatures as high as 20°C were observed during the summer (Dawson & Morin 1996). At Nanisivik Mine, results indicated that a 2-m cover consisting of natural cover materials (shales, till and sand) without any sulfide minerals will adequately induce the permafrost table to move up, thereby maintaining reactive tailings below 0°C on an annual basis (Fig. 7). However, as the test pads in Nanisivik during the test period were almost snow-free, additional snow in larger covered areas may result in increased insulation during autumn and higher early winter temperatures, but later thaw onset in spring. The impact of snow on ground temperatures and AMD generation in general is discussed in greater detail in section 5.3.

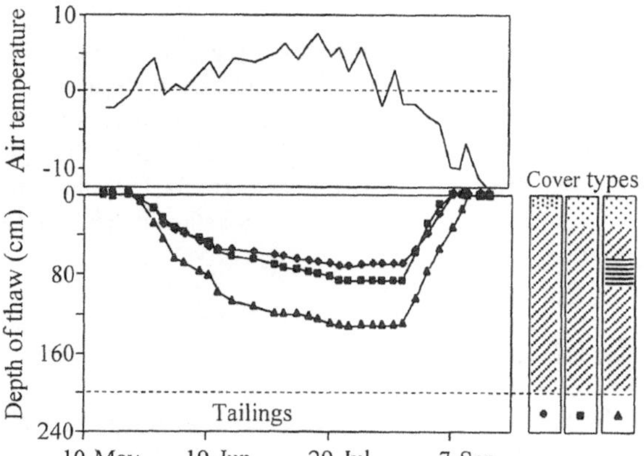

Figure 7. Active layer depths in tailings 1997 covered by two metres of inert covers near Nanisivik Mine, Canada (unpublished data from Nanisivik Mine). Covers consist of sand and gravel (dots), till (horizontal lines), and shale (hatched lines).

As shown in Figure 2, temperatures slightly below zero will not guarantee negligible sulfide oxidation or completely frozen tailing conditions. Therefore, the temperature at the tailing-cover interface becomes an important environmental parameter controlling AMD generation. Temperature readings below various cover schemes in Nanisivik reveal that temperatures typically vary between -15 to few degrees below zero. Thus, the recommended thickness of covers must be based on predictions of the sub-zero temperatures below the cover and the acceptance of the corresponding AMD generation. Several models are available for such predictions. Kyhn & Elberling (2001) calibrated the WinSoil model (Jansson 1998) to observed field conditions in Nanisivik and simulated the temperature below different cover schemes. Figure 8A shows an example of the importance of cover thickness on the temperatures at the tailing-cover interface. In this case snow was not considered.

Although freezing is considered a method to control AMD generation in permafrost areas, several factors complicate the application of ground temperature as a fine-tuned control method for AMD generation. As pyrite oxidation leads to a heat generation of approximately 12,000 kJ per kg pyrite (Dawson & Morin 1996) the oxidation process may influence the rate of ground freezing and result in higher temperatures near sulfide minerals than the temperature measured in bulk material. Assuming an annual average uptake of about 38 moles O_2/m^2 as observed in Nanisivik 1998/99 (Elberling 2000) more than 1 kg pyrite per m^2 oxidizes annually, generating about 32 kJ/m^2 per day. Compared to 334 kJ required to melt 1 kg of ice (Godwaldt et al. 1999) the rate of heat loss from oxidizing tailings to the surroundings may be enough to maintain elevated temperatures and AMD generation for some time after bulk freezing.

Further laboratory studies of bacterial adaptation and temperature feedback mechanisms on sulfide oxidation in the temperature range 0 to -10°C seem crucial in order to provide reasonable long-term estimates for AMD generation at sub-zero temperatures.

5.2 *Engineered dry and partly-saturated covers*

Engineered layered soil covers have been used to achieve an effective reduction of oxygen and water movement within tailings. Typically, cover systems are constructed from natural sediments and consist of layers with different hydraulic properties. The upper layers aim to prevent erosion and evaporation, the middle layers aim to retain water and act as a barrier for oxygen, while the lower layers aim to prevent drainage from overlying layers. Using numerical simulation, Akindunni et al. (1991) showed that such barrier systems are possible when taking into account the air-entry value of the middle layer and the pressure head at which the lower layer approaches residual saturation. However, in permafrost areas cracking caused by seasonal freezing and reduced moisture caused by evaporation may alter the transport properties and even enhance oxygen and water migration due to preferential flow in crack systems.

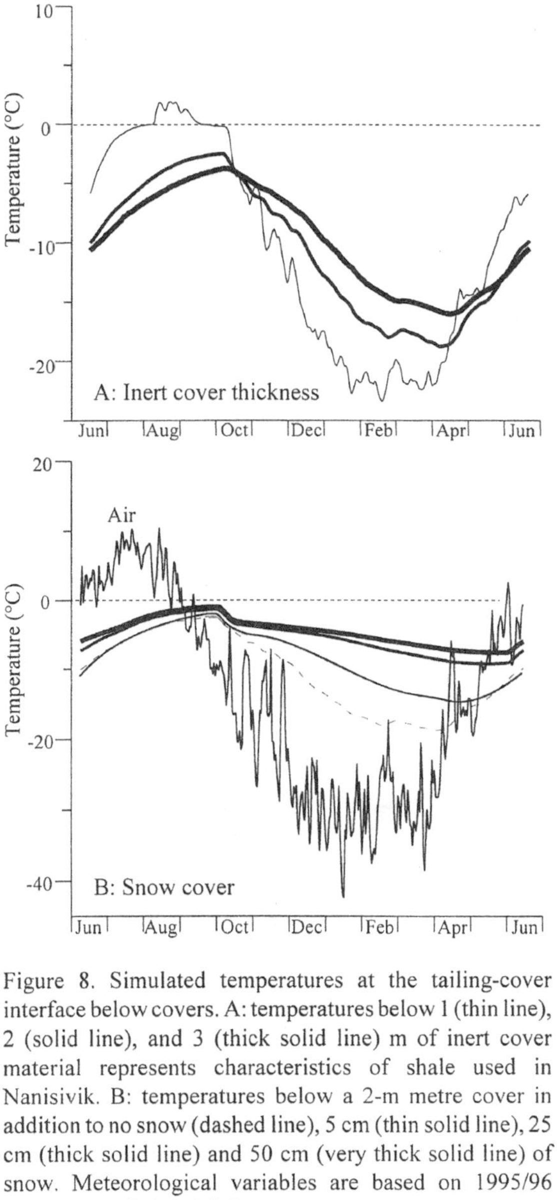

Figure 8. Simulated temperatures at the tailing-cover interface below covers. A: temperatures below 1 (thin line), 2 (solid line), and 3 (thick solid line) m of inert cover material represents characteristics of shale used in Nanisivik. B: temperatures below a 2-m metre cover in addition to no snow (dashed line), 5 cm (thin solid line), 25 cm (thick solid line) and 50 cm (very thick solid line) of snow. Meteorological variables are based on 1995/96 observations in Nanisivik.

5.3 *Snow fences and snow covers*

In the Arctic, most precipitation falls as snow during winter. Snow covers are known to prevent wind erosion and redistribution of mine tailings during storms. Consequently, snow fences have been built, resulting in additional snow accumulation in tailing areas. The snow cover may also limit oxygen diffusion during winter and delay thawing in spring, thereby limiting the period of active layer development. However, decades of field observation in Alaska and Canada have shown that snow has a crucial influence on ground temperatures because its insulating effects that reduce winter heat loss (e.g., Nicholson & Granberg 1973). This insulation effect again depends on the thickness and density of the snow layer and it has been concluded that ground temperature variations from year to year are closely related to annual variations in snow accumulation (Nicholson & Granberg 1973). Results presented by Brooks et al. (1997, 1998) confirm that snow covers insulate frozen ground from the low winter air temperatures resulting in surprisingly high subnivial levels of bacterial activity under seasonal snow cover. In a study by Brooks et al. (1997) fluxes of CO_2 and N_2O as well as active microbial biomass were measured when soil temperatures under a snow cover ranged from -5°C to 0°C. These observations led the authors to suggest that subnivial bacteria activity may be a significant component of the annual C and N cycling in these environments. Elberling (2000) and Kyhn & Elberling (2001) have suggested that similar conditions exist in sulfidic tailings deposited in permafrost regions. The study by Kyhn & Elberling (2001) presented simulations based on WinSoil to verify the potential insulation effect of a snow cover (Fig. 8B). Further investigations are needed to quantify the effect of snow as a barrier for oxygen gas compared to snow as a barrier for heat controlling ground temperatures.

Although global changes in air temperatures have received much attention, global and local changes in snow covering (winter precipitation and wind regimes) may prove to be the most critical parameter for long-term AMD generation. This is probably particularly important for facilities constructed for permanent frozen disposal of well-drained tailings whereas facilities for permanent subaqueous disposal will be more sensitive to global changes that affect the water balance, such as summer evaporation rates and annual precipitation.

5.4 *Subaqueous tailing disposal*

Mine tailings have been placed under water in tailing impoundments in many locations worldwide, primarily to minimize oxygen diffusion into the tailing materials (Fig. 3). This disposal strategy has also been used in permafrost areas and may be attractive in areas where mine tailings can be placed in natural water bodies in a controlled fashion as is the case at Polaris Mine, a lead-zinc mine on Little Cornwallis Island, Canada. Here, mine tailings have been discharged 26 m below the lake surface (Dawson & Morin 1996).

In permafrost areas without a suitable natural water body, construction of dams may be an alternative. The effect on AMD of covering tailings with water in permafrost areas has not received much attention although the potential effect is well known and can be easily illustrated. Considering the maximum oxygen diffusion in tailings covered with 30 cm saturated sand again covered by a thin cover of oxygen-saturated water. The maximum amount of oxygen which at steady state can diffuse through the 30 cm saturated cover at 5°C will be approximately 1.5×10^{-4} mol/m^2 per day (Eq. 4) or about 10,000 times less than the oxygen uptake observed in dry tailings at 5°C in Nanisivik (assuming instantaneous oxygen uptake at the tailing surface). The above predictions are consistent with laboratory tests on the reactivity of tailings under water (Mugo et al. 1999, Elberling & Damgaard 2001), indicating that the storage of reactive tailings under water is effective in attenuating long-term fluxes of contaminants to the overlying water column.

The local hydrology and water balance are important factors to consider when designing a water-covered tailing impoundment. After mine closure, the water balance will be based on the precipitation, melt water input and evaporation of the catchment area. Strategies can be designed to maintain an equilibrium state, such as snow fences to increase snow accumulation and meltwater and construction of channels to regulate water input and output to the pond area. Difficulties include sufficient storage capacity for extreme precipitation and the potential stability of dyke constructions in permafrost-affected areas.

Increasing oxygen solubility in cold water has been suggested as a concern when designing water covers (Dawson & Morin 1996, Davé & Blanchette 1999). The amount of oxygen in the water increases by a factor of 1.4 as the temperature decreases from 15 to 0°C, whereas the oxygen diffusivity decreases by a factor of 1.6 (Ramsing & Gundersen 1994) which is seldom taken into account. Comparisons of a saturated system at two different temperatures, using the above example with 30 cm saturated sand, the combined effect of increasing oxygen solubility and decreasing diffusivity will result in a reduction of the potential flux of oxygen at 2°C to approximately 90% compared to the flux at 15°C. The results of such simple calculations are consistent with measurements of micro-scale oxygen gradients and diffusivity in saturated tailings investigated at various temperatures (Elberling & Damgaard 2001). It can be concluded that subaqueous tailings disposal is probably a more effective remediation action in cold regions than elsewhere.

6 CONCLUSIONS AND FUTURE RESEARCH

There is currently great interest in understanding the chemical and physical processes occurring in frozen sediments in both natural and engineered systems. Concern about contamination in the Arctic is growing and frozen barriers have recently been considered as an option for preventing AMD generation. This review shows that AMD is not eliminated nor reduced to negligible levels at sub-zero temperatures, although it is reduced significantly. Very low temperatures for most of the year in the continuous permafrost zone can be shown to inhibit AMD generation. The effect of using ground temperature and freezing as the major control on tailing oxidation and release of pollutants in permafrost areas is only partly understood. Reduction of AMD may also be achieved by limiting the availability of oxygen to the tailings to maintain the reactive material nearly or fully water-saturated. In the latter case, natural freezing and the permafrost provide additional protection against AMD generation and reduce contaminant movement.

Several key issues need to be better understood for accurate prediction of acid mine drainage in cold regions and the overall limiting effects of freezing. These issues include environmental controls on oxidation processes below 0°C, the lower temperature limit for sulfide oxidation and the behaviour of contaminants migration in aqueous solutions in unsaturated and saturated frozen soils.

ACKNOWLEDGEMENTS

Funding for this review was provided by the Danish Environmental Protection Agency, Ministry of Environment and Energy, Denmark, as part of of the MIKA project "Chemical fluxes in frozen soil". MIKA is now known as DANCEA (Danish Cooperation for Environment in the Arctic). The author is solely responsible for the results and conclusions presented. Special thanks go to B.H. Jakobsen, G. Asmund, R. S. Sletten and K. LeDrew for comments on the manuscript.

REFERENCES

Akindunni, F., R.W. Gillham & R.V. Nicholson 1990. Numerical simulations to investigate moisture-retention characteristics in the design of oxygen-limiting covers for reactive mine tailings. *Can Geotech. J.* 28: 446-451.

Barnall, D. & D. Slotfeld-Ellington 1983. Pulsed nuclear resonancestudies of doped ice. *J. Phys. Chem.* 87: 4321-4329.

Black, P.B. & M.J. Hardenberg 1991. Historical perspectives in frost heave research: the early works of S.Taber & G. Beskow. *USA Cold Regions Research and Engineering Laboratory*, Special Report 91-23.

Boyd, W.L. & J.W. Boyd 1964. The presence of bacteria in permafrost of the Alaskan arctic. *Can. J. Microbiol. 102*: 917-919

Brooks, P.D., S.K. Schmidt & M.W. Williams 1997. Winter production of CO_2 and N_2O from alpine tundra: environmental controls and relationship to inter-system C and N fluxes. *Oecologia*, 110: 403-413.

Brooks, P.D., M.W. Williams & S.K. Schmidt 1998. Inorganic nitrogen and microbial biomass dynamics before and during spring snowmelt. *Biogeochemistry* 43: 1-15.

Cameron, E.M. 1977. Geochemical dispersion in mineralized soils of a permafrost environment. J. Geotechn. *Exploration*. 7: 301-326.

Cameron, E.M. 1979. Effect of graphite on the enhancement of surficial geochemical anomalies originating from the oxidation of sulphides. J. Geotechn. *Exploration*. 12: 35-43.

Dawson, R.F. & K.A. Morin 1996. Acid mine frainage in permafrost regions: issues, control strategies and research requirements, *Department of Indian and Northern Affairs Canada*, MEND report 1.61.21. CANMET, Natural Resources.

Elberling, B. 1998. Processes controlling oxygen uptake rates in frozen mine tailings in the Arctic. In: *Ice in Surface Waters* (H.T.Shen, ed), 183-188. A.A.Balkema Rotterdam/Brookfield.

Elberling, B. 2000. Environmental controls of the seasonal variation in oxygen uptake in sulfidic tailings deposited in a permafrost-affected area. *Water Resour. Res*. (in press).

Elberling, B. & L.R. Damgaard 2001. Micro-scale measurements of oxygen diffusion and consumption in subaqueous sulfide tailings. *Geochim. Cosmochim. Acta* (in press).

Elberling, B. & B.R. Langdahl 1998. Natural heavy-metal release by sulphide oxidation in the High Arctic. *Can. Geotech. J*. 35: 895-901.

Elberling, B. & R.V. Nicholson 1996. Field measurements of oxygen diffusion and sulphide oxidation in mine tailings. *Water Resour. Res*. 326: 1773-1784.

Elberling, B., R.V. Nicholson & D.V. Darren 1993. Field evaluation of sulphide oxidation rates. *Nordic Hydrol*. 245: 323-338.

Elberling, B., R.V. Nicholson & J.M. Scharer 1994. A combined kinetic and diffusion model for pyrite oxidation in tailings: a change in control with time. *J. Hydrol*. 157 47-60.

Elberling, B. A. Schippers & W. Sand 1999. Bacterial and chemical oxidation of pyritic mine tailings at low temperatures. *J. Cont. Hydrol*. 41: 225-238.

Evangelou, V.P. 1995. Pyrite oxidation and its control. *CRC Press*, Inc. 293 pp.

Forster, J. 1887. Ueber einige Eigenschaften leuchtender Bakterien. *Zentr. Bakteriol. Parasitenk. Infekt. Hyg*. 2:337-340.

Gilichinsky, D.A., E.A. Vorobyova, L.G. Erokhina, D.G. Fyodorov-Davydov & N.R. Chaikovskaya 1992. Long-term preservation of microbial ecosytems in permafrost. *Adv. Space Res*., 12:255-263.

Godwaldt, R.C., K.W. Biggar & D.C. Sego 1999. AMD generation at sub-zero temperatures. In: *Assessment and Remediation of Contaminated sites in Arctic and Cold Climates*, 75-82. ARCSACC'99, Edmonton, May 3-4, 1999.

Hallet, B. 1978. Solute redistribution in freezing ground. In proceedings of the *Third International Conference on Permafrost*, 85-91. Edmonton, Canada, 1978.

Jansson, P.E. 1998. Simulation model for soil water and heat conditions: description of the SOIL model. *Commun*. 98:2, Swedish University of Agricultural Science, Department of Soil Science, Uppsala, Sweden.

Jaynes, D.B., A.S. Rogowski & H.B. Pionke 1984. Acid mine Drainage from reclaimed coal strip mines 1. Model description. *Water Resour. Res*. 20: 233-242.

Kalin, M. 1987. Ecological engineering for gold and base metal mining operations in the North West Territories, in *Northern Affairs Program*, Department of Indian Affairs and Northern Development, pp. 1-64, Environmental Studies no. 59.

Kjøller, A. & S. Ødum 1971. Evidence for longevity of seeds and microorganisms in permafrost. *Arctic*. 243: 230-232.

Konrad, J.M. & A.W. McCammon 1990. Solute partitioning in freezing soils. *Can. Geotech. J.* 67: 726-736.

Kwong, Y.T.J. 1995. Geochemical, electrochemical and microbial control of sulphide oxidation and their implication for the abatement of acid drainage. In: *International Workshop on Abatement of Geogenic Acidification in Mining Lakes*, Sept. 4-6, 1995, Magdeburg, Germany.

Kyhn, C. & B. Elberling 2001. Frozen cover actions limiting AMD from mine waste deposited on land in Arctic Canada. *Cold Region Science and Technology* (in press).

Langdahl, B.R. 1999. Acidophilic microorganisms in a High Arctic gossan environment. PhD. Thesis, *Department of Microbial Ecology, Institute of Biological Sciences. University of Aarhus*, Denmark. 241 pp.

Langdahl, B.R & K. Ingvorsen 1997. Temperature characteristics of bacterial iron solubilisation and ^{14}C assimilation in naturally exposed sulfide ore material at Citronen Fjord, North Greenland (83°N). *FEMS Microbiol. Ecol.* 23:275-283.

Marion, G.M. 1995. Freeze-thaw processes and soil chemistry. CRREL special report 95-123. US Army Corps of Engineers. *USA Cold Regions Research and Engineering Laboratory*, 1-28.

Melak, J.M. & J.O. Sickman 1995. Snowmelt induced chemical changes in seven strems in the Sierra Nevada, California. *Biogeochemistry of seasonally snow-covered Catchments* (Proceedings of a Boulder Symposium, 221-234. July 1995. IAHS Publ. 228.

Meldrum, J.L., H.E. Jamieson & L.D. Dyke 1999. Laboratory determination of sulphide oxidation potential in permafrost using tailings from Rankin Inlet, Nunavut. In: *Mining and the Environment* 2, 119-126. Sudbury, Canada. September 12-16, 1999.

Mironenko, M.V., S.A. Grant, G.M. Marion & R.E. Farren 1997. FREZCHEM2. A chemical thermodynamic model for electrolyte solutions at subzero temperatures, pp. 1-40, *Cold Regions Research & Engineering Laboratory*, CRREL Report 97-5.

Morita, R.Y. 1975. Psychrophilic bacteria. *Bacteriol. Rev.* 39: 144-167.

Mugo, R.K., D. McDonald & G.W. Poling 1999. Subaqueous tailings disposal in freshwater and marine environments - results of predictive geochemical testing using tailings with different compositions. In: *Mining and the Environment* 2, 99-108. Sudbury, Canada, September 12-16, 1999.

Nicholson, F. H. & H.B. Granberg 1973. Permafrost and snow cover relationships near Schefferville. *Second International conference on Permafrost*. 151-158. Yakutsk, U.S.S.R., 13-28 July 1973.

Nicholson, R.V., R.W. Gillham & E.J. Reardon 1988. Pyrite oxidation in carbonate-buffered solution: 1 experimental kinetics. *Geochim. Cosmochim. Acta*. 52: 1077-1085.

Nordstrom, D.K. & G. Southam 1997. Geomicrobiology of sulfide mineral oxidation. Chapter 11 in *Geomicrobiology*. Mineralogical Society of America. Reviews in Mineralogy. 35: 362-390.

Pantelis, G. & A.I.M. Ritchie 1992. Macroscopic transport mechanisms as a rate limiting factor in dump leaching of pyritic ores. *Appl. Mat. Model.* 15: 136-143.

Paktunc, A.D. 1999. Characterization of mine wastes for prediction of acid mine drainage. In: *Environmental impacts of mining activity*. J. M. Azcue (Ed.). Springer, 19-40.

Price, W.S., H. Ide & Y. Arata 1999. Selfdiffusion of supercooled water to 238 Kusing PGSE NMR diffusion measurements. *J. Phys. Chem.* 103: 448-450.

Ramsing, N. & J. Gundersen 1994. Seawater and gases. Tabulated physical parameters, version 1.0. Unisense A/S, Århus, Denmark.

Rasch, M, B. Elberling, B.H. Jakobsen & B. Hasholt 2000. High-resolution measurements of water discharge, sediment, and solute transport in the river Zackenbergelven, Northeast Greenland. *Arctic, Antarctic, and Alpine Research* 32: 336-345

Sheeran, D.E. & R.N. Yong 1975. Water and slat redistribution in freezing soils. In *Conference on Soil Water Problems in Cold Regions*, 58-69. Division of Hydrology, AGU, Calgary, Alberta, Canada, May 6-7, 1975.

Troeh, R.F., J.D. Jabro & D. Kirkham 1982. Gaseous diffusion equations for porous materials. *Geoderma*, 27: 239-253.

Mineral exploration and exploitation

Mining in the Arctic, Olsen, Lorentzen & Rendal (eds), © 2001 Swets & Zeitlinger, Lisse, ISBN 90 5809 177 5

Mining Concept for the Citronen Fjord Zinc Deposit, NE Greenland

Erik O. Andersen
Platinova A/S, 181 University Ave, Ste 1414, Toronto, ON Canada M5H 3M7

ABSTRACT: The Citronen Fjord "sedex" massive sulphide zinc deposit was discovered by Platinova A/S in 1993. It is the world's northernmost commercial project, located at 83.1° N and 28.4° W directly on tide water. Drilling located potentially economic zones of which the Beach Zone is the best defined. The Beach Zone is shale hosted with a diluted resource of 8 million tons grading 7.1% Zn and 0.9% Pb. The zone is flat lying on two horizons separated by 5 m at 100 m below surface and extending laterally 3000 m by 400 m. Mining is proposed with a roadheader in a drift-and-rib-pillar method utilizing conveyors for haulage to the mill. Metallurgical testing has confirmed that a combination flotation and leach circuit will give maximum recovery. A detailed navigation study concluded that a ship like the MV Arctic could have an average navigation season of about 60 days.

1 INTRODUCTION

The northern part of the Canadian Arctic Islands and Greenland are underlain by the Paleozoic Franklinian Basin. This Basin has long been the target of exploration for zinc-lead mineralization over the range of geologic models associated with sedimentary basins. The discovery of the Polaris deposit in 1971 by Cominco Ltd. underscored the economic potential of the basin.

Two other zinc-rich deposits, the Nanisivik on Baffin Island and the Black Angel in west central Greenland are also sedimentary hosted. Even though the host formations for these deposits are older than the Franklinian Basin, they add attraction to the zinc-rich nature of north Greenland.

The Red Dog zinc-lead-silver deposit in Alaska demonstrates just how big and alluring sedex type deposits can be. Red Dog reserves stand at 142 million tons of 15.8% zinc, 4.3% lead and 83 grams per ton silver - and growing.

A mostly unfruitful reconnaissance expedition in northwest Greenland in 1992 was conducted by a Platinova-Nanisivik Mines Ltd. prospecting team. The following year, a scaled down reconnaissance programme was mounted by Platinova alone to prospect northeast Greenland. Although economic geological information in this part of Greenland was sparse, full use was made of clues provided by the prior work of the Geological Survey of Denmark and Greenland (GEUS) (van der Stijl & Mosher 1998).

Spectacular gossan zones anomalous in zinc were discovered on the first day of prospecting. This discovery was followed up immediately with a diamond drill programme. As evidence mounted that there was a major zinc deposit in the Citronen Fjord area, drill programmes were carried out each year from 1993 to 1997. In total, 32 400 m of diamond drilling was completed in 143 holes.

2 LOCATION AND LOGISTICS

The Citronen Fjord project is the world's northernmost land-based commercial project ever and is likely to remain so, being only 50 km south of the world's northernmost point of land. Citronen Fjord is a small bay on the south side of Frederick E. Hyde Fjord in Peary Land, northeast Greenland. The actual location is 83.1 N and 28.4 W. Of critical importance to the project has been the existence of the Danish military base at Station Nord, 200 km to the SE. The cooperation and assistance of the base personnel and the Danish military, in general, cannot be overstated. The airstrip at Station Nord has allowed Platinova to operate large aircraft including, Boeing 727's, Hercules', and HS 748's into the region. Unfortunately, Station Nord is only serviced by air - there is no port possibility at the base due to shallow offshore waters.

Svalbard lies about 900 km ESE of Citronen Fjord. The airport at Longyearbyen is the closest civilian airport to the site and offers daily jet service

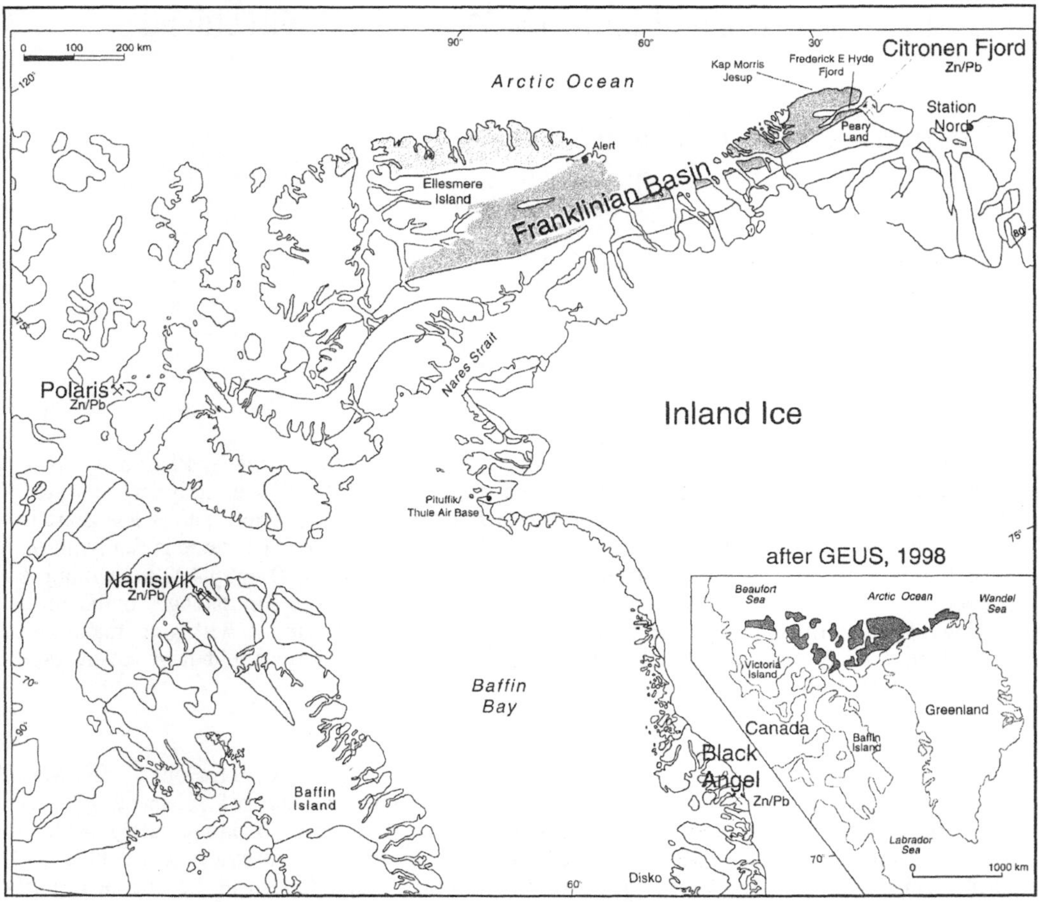

Figure 1. Franklinian Basin Location Map

from Norway. Svalbard is accessible by open water shipping year-round.

Frederick E. Hyde Fjord is uncharted, but believed to be quite deep, even at its mouth where limited soundings indicate minimum water depths fo 150 m. The fjord is very wide all the way in from the open sea only narrowing to 7.5 km at Citronen Fjord. In addition, the ice in the fjord clears nearly every year, at least west of Citronen Fjord, and always offers wide summertime leads east of Citronen Fjord to the sea.

Citronen Fjord, itself is more of a bay than a fjord. It is located on the south side of Frederick E. Hyde Fjord, about 50 km from its mouth. Citronen Fjord is only 4 km long. It has been well sounded, showing water depths mostly in excess of 50 m. Along the southwest side there is deep water in excess of 30 m right up to a steep sided shore line that offers an excellent harbour location. Tides within Citronen Fjord are less than 0.5 m.

Figure 7, at the end of this paper, shows a general view over the Beach Zone towards Citronen Fjord.

3 CLIMATE AND PHYSIOGRAPHY

The climate in the region is not nearly as harsh as one might assume. Summer temperatures average about +3 C and can reach as high as +15 C. Winter temperatures average about -30 C and probably seldom drop below -40 C.

Wind speeds have been recorded as high as 25-30 knots during the summer, although there can be periods extending up to two weeks of calm. Wind speeds during winter are not well known, although storms are probably more common.

Precipitation is very low, mostly as snow, in the order of 10 to 20 cm annually.

Being in the high arctic, the extremes of winter darkness and summer light are marked. Total darkness lasts from mid-October until late February. Total daylight lasts from mid-April to early September.

The ice on Citronen Fjord is completely gone by late July and begins to form again in early September.

An important factor for considering an approach to mining at Citronen Fjord is that the entire area of mineralization is within permafrost. Drilling has confirmed permafrost to depths of 200 m.

Another important factor for considering mining techniques is that the region is within a seismically active zone. Tremors, perhaps as high magnitude 4 or 5, have been observed by field personnel at Citronen Fjord. This is about the maximum expected for the area.

4 GEOLOGICAL SETTING

The Citronen Fjord mineralization occurs within the Ordovician Amundsen Land Group. The Beach Zone, which is the focus of this paper, occurs in the lower portion of the Footwall Shale.

The geological setting is fully described by van der Stijl & Mosher (1998). Four separate zones of mineralization have been identified: the Discovery, Beach, Esrum and West Gossan. This paper deals only with the best known zone, the Beach Zone.

The mineralization of the area is of the sedimentary exhalative (sedex) type. That is, a hydrothermal source provided zinc-rich fluids that reached the sea floor of a secondary shale basin through a system of regional and local faults. These fluids formed zinc and iron-rich sulphides and precipitated onto the sea floor upon escape from sea-floor fissures. The quiet-water sulphide precipitation occurred contemporaneously with silt sedimentation. Sometimes silt sedimentation proceeded at rates faster than sulphide sedimentation resulting in a banded cross-section ranging from millimetre to metre scale. The result, unfortunately, is that silt layers form an internal dilution within the zinc-rich section.

From time to time, a catastrophic debris flow would completely cover the sulphide-silt sediments. The debris flows resulted from de-stabilizing events on the steeper margins of the basin that sent unsorted blocks of carbonate and mudstone sliding down the slope into the basin centre. A debris flow event was followed by another period of quiet sedimentation during which the pre-existing plumbing system to surface re-established itself, off-set somewhat from

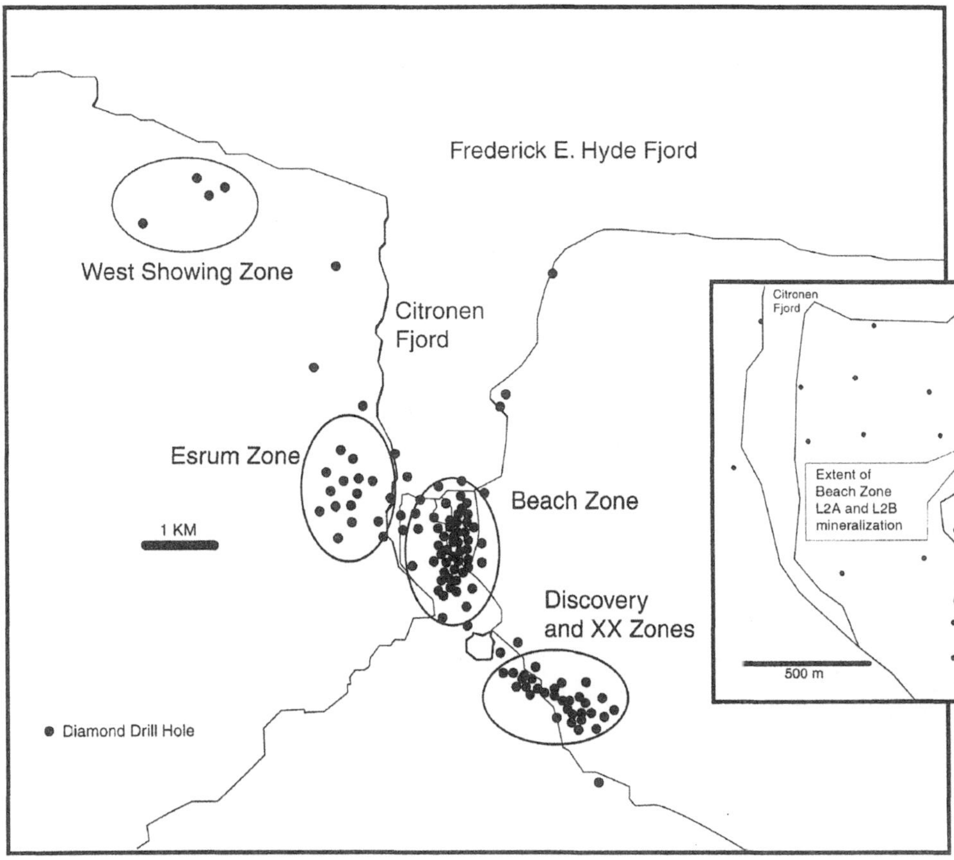

Figure 2. Citronen Fjord Mineralized Zones Location Map

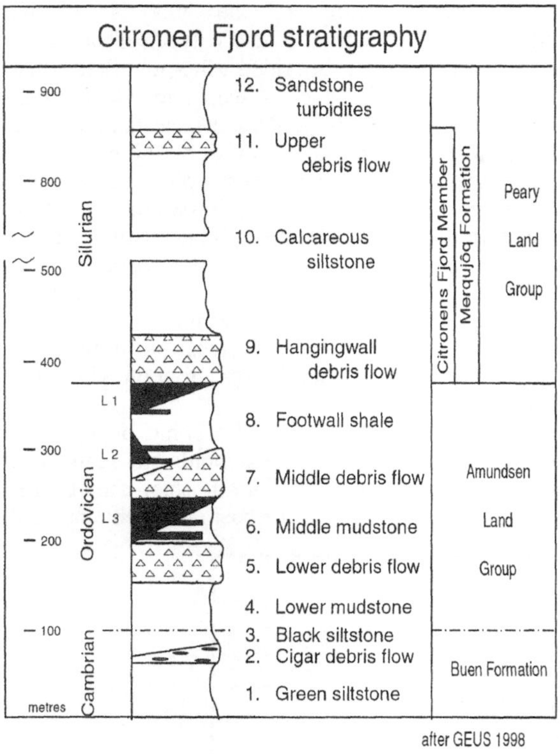

Figure 3. Schematic stratigraphic column in Citronen Fjord region

the previous route.

Repeated cycles of this process produced at least three identifiable mineralized zones or levels at Citronen Fjord. The Beach Zone is stratigraphically below the Discovery Zone and above the Esrum Zone. The location of the Zones is shown in Figure 2 which also shows the density of diamond drilling on the property. Figure 3 shows the regional stratigraphy including the relationships of the mineralized zones.

5 MINERALIZATION

The mineralization of economic interest consists of massive and bedded pyrite with variable amounts of sphalerite and minor galena. There are no other minerals of economic potential. The sulphides are generally fine- to medium-grained; some are weakly bedded and laminated, others lack sedimentary features. Bedded and laminated sulphides are the most important from an economic point of view. There are also significant areas that show dendritic textures, likely due to re-mobilization. Individual layers range in thickness from 1 mm to 1 m, although decimetre thicknesses are most common. Beds in excess of 30 cm are rare. Cycles of beds form the levels of mineralization that can have combined thicknesses of up to 20 m.

Sulphides contain clay and silt as impurities and mudstone commonly contains up to 20 per cent

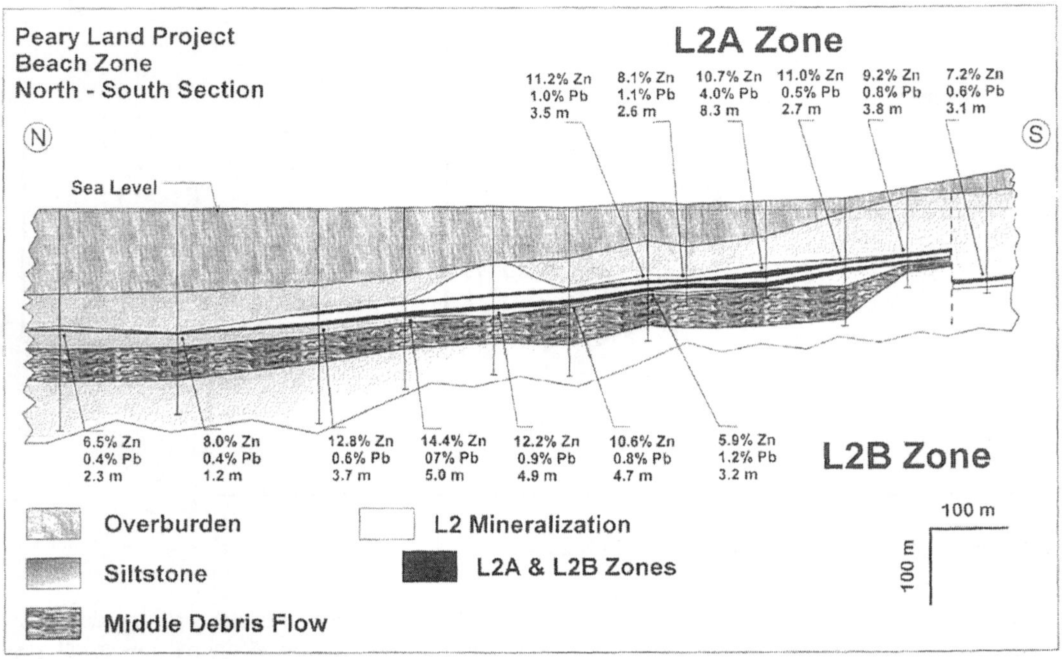

Figure 4. Longitudinal, north-south, cross section through the Beach L2A and L2B Zones

disseminated pyrite. It is nearly always difficult to visually identify sphalerite within the pyrite beds. A portable XRF analyser was used on site to obtain immediate estimates of zinc grade. The XRF results were vital as a guide to drill programme strategy. Pyrite often occurs without sphalerite but sphalerite never occurs without pyrite.

The mineralized beds are grouped into "levels" of mineralization. A level represents a time period of zinc sulphide-rich sedimentation. The result is a zone with fairly definitive upper and lower high-zinc assay boundaries. The Beach Zone is within a general time period called Level 2 or, simply, L2. A closer examination of the mineralization in the Beach Zone reveals that there is an upper and a lower level separated by about 4 to 5 m and offset somewhat from each other. The upper mineralization is designated L2A and the lower L2B. Figure 4 is a longitudinal section showing the detailed relationships in the L2 zone.

6 MINERAL RESOURCES

Platinova calculated an unclassified, undiluted resource for the L2A and L2B zones totalling 6.75 million metric tons grading 8.64% Zn and 1.10% Pb using a cutoff of 6% Zn over 2 m (Platinova A/S 1997).

The mineralized zones have an average dip of 7 degrees to the north except at the south end where the dip is about 10 degrees. The height of the mineralization varies from 2 m to 20 m.

The mineralization is 85 m below surface at the southern end and 160m m at the northern extent. The hanging wall of both zones is a laminated mudstone-siltstone. The northern end of the mineralization is covered by an unconsolidated, although frozen, overburden to depths of 120 m, but not coming closer than 30 m to the top of the L2A Zone.

7 PRODUCTION CONCEPT DEVELOPMENT

By the end of 1997, the identified mineral resource at Citronen Fjord was approaching a threshold at which it was possible to develop production concepts and order-of-magnitude cost models.

It was recognized that the resources in the Beach Zone itself may not be sufficient to support the construction of a mine. However, in order to provide an economic basis for continued exploration at Citronen Fjord, especially in the Esrum and West Gossan Zones, it was timely to document mining and processing concepts that could be applied to production from the deposit.

Considering the grade of the known resource and the remote arctic location of the project, the following objectives were set for a production concept:

1. The operation should be as mechanized as possible, keeping staffing levels to a minimum;
2. Materials handling should be as efficient as possible and make use of technology to reduce the number of equipment operators;
3. A mining method should be selected to keep controllable and controllable dilution to a minimum;
4. The mining method should allow for maximum extraction of the resource;
5. The processing circuit should remove the internal non-sulphide silt particles as early in the process as possible;
6. A process should be selected that will yield the highest value per ton of product leaving the mine gate;
7. The operation should be as physically compact as possible.

Kvaerner Metals, a division of Kvaerner Canada Inc. was engaged to prepare the Order-of-Magnitude Study in 1998. Close cooperation between Kvaerner and Platinova ensured that alternatives were evaluated at early stages. Only sufficient work was done to determine whether or not a particular alternative had merit. Ideas or concepts that did not appear to be optimal were rejected quickly in order to limit the scope of the final report.

8 UNDERGROUND MINING

It was decided that full advantage should be made of the fact that the mineralization is in permafrost. There is concern that the laminar nature of the hanging wall would create a weak mining back susceptible to overbreak. Any overbreak would negatively affect both dilution and extraction and increase the risk of falls of ground.

The first conclusion was, therefore, that permafrost conditions should be preserved within the mine. Preserving permafrost implies that water should be kept out of the mine operations. The use of diesel equipment underground should also be minimized or eliminated to reduce heat production.

Since dry drilling is proven technology at both the Polaris and Nanisivik mines in the Canadian NWT, this was considered the obvious choice if conventional drill jumbos were to be used.

The objective to reduce the use of diesel equipment also led to the conclusion that, as much as possible, all-electric mobile equipment should be used. The choice of electric equipment would also have the added benefit of reducing ventilation requirements.

The depth to the mineralization precluded consideration of open pit mining. However, the depth to the top of the mineralization of only 85 m, at the shallowest, made the choice of access by decline obvious. Shaft access was, therefore, not considered.

Initially, a conventional room-and-pillar technique for mining was considered. Drilling would be by jumbo, probably two-boom, mucking with 2.5 to 6 cu yd LHD's and haulage by truck up the ramp to the mill.

A number of problems were recognized with a conventional mining approach. Of major concern was the back height required to operate even the lowest profile equipment available. Significant parts of the mineralization are only 2 m thick. Operating equipment requiring more headroom than this would result in excessive dilution, perhaps as much as 100% and/or low productivity rates.

Even with conventional mining, conveyor transport to surface of run-of-mine muck was considered desirable since this would eliminate the use of underground trucks and truck drivers.

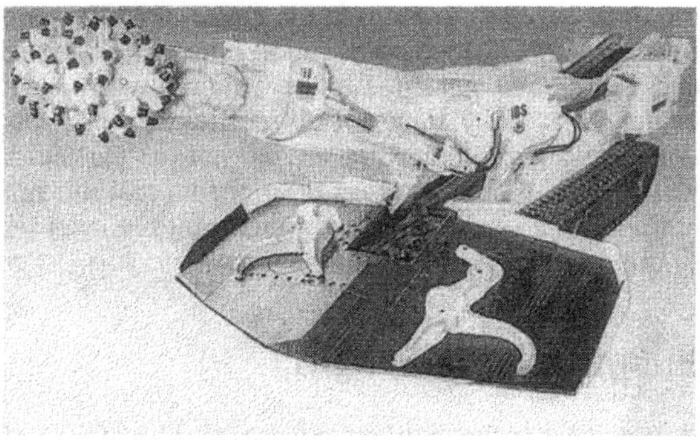

Figure 5. Typical roadheader showing cutting head and gathering arms.

During the course of considering mining method options, the possibility of using continuous mining or tunnelling machines arose. Further investigation led to the conclusion that, so called, roadheaders normally used in tunnelling might find an ideal application in the Beach Zone.

Critical to deciding if roadheaders could be used is the determination of the compressive strength of the ore. Unfortunately, a sample of sufficient size to make this determination was not available. However, visual examination of drill core samples indicates that the marked lamination and poor consolidation of the sulphides and mudstones will very likely make the ore amenable to efficient mining by roadheader. An average compressive strength of 35 mPa was assumed by Kvaerner for the study.

The equipment choice recommended by Kvaerner is the ABM330 Alpine Boom Miner, which can drive a heading 6 m wide by 2.5 m high. Continuous mining technology is developing rapidly and it is expected that other choices will be available when actual choices for Citronen Fjord will have to be made. Figure 5 shows a typical roadheader.

The advantages of using continuous mining equipment like a roadheader include:

1. Smooth breakage to stope backs, greatly contributing to back stability, particularly when compared to conventional drill and blast techniques;
2. Excellent control of dilution, especially considering the low operating heights;
3. Relative fine size of run-of-mine muck reducing crushing requirements;
4. When combined with a self-propelled transfer conveyor and a central main conveyor to surface, there is the opportunity to haul muck from the mining face right to the primary crusher on surface without the use of any mobile equipment;
5. Reduced operating staff levels;
6. No blasting cycle, therefore no waiting between shifts for smoke and dust to clear the mine.

With these advantages, it was decided not to consider conventional mining methods further for this study.

In addition to use in the production mode, the roadheaders would be used to drive the access decline from surface to the stoping areas. The access ramp is proposed to be 6.0 m wide by 3.0 m high at -13.2% grade in one straight drive. A straight drive will facilitate the installation of a main haulage conveyor beside an access roadway.

8.1 *Stoping Method*

Once it had been established that the equipment choice would be continuous mining roadheaders, a stoping and extraction plan had to be developed to fit the operating characteristics implied in the equipment choice.

A herring-bone patterned, "drift-and-rib-pillar" mining method is proposed. The first step is the driving of a central access drift in the long axis of the ore (north to south). From this access drift, the roadheader will drive drifts in the ore 6.0 m wide and to the height of the ore, but a minimum of 2.5 m. More than one pass may be required in areas of thicker ore. Figure 6 illustrates schematically the drift-and-rib-pillar stoping layout.

Extendable, mobile transfer conveyors will take the ore directly from the roadheader to the main conveyor. Side-by-side production drifts will have a 4.0 m wide rib pillar left between them.

As a production drift is completed, it will be filled with tailings which will be allowed to freeze. Once the fill has frozen, a second pass with the roadheader will mine the remaining rib pillar between the first pass tailings-filled drifts.

It is expected that the roadheader mining using the "drift-and-rib-pillar" method will yield 95% extraction of the ore and limit dilution to 25%. Applying these factors to the undiluted mineral resource calculated for the Beach Zone, the diluted, extractable resource is 8.0 million tons at 7.11% Zn

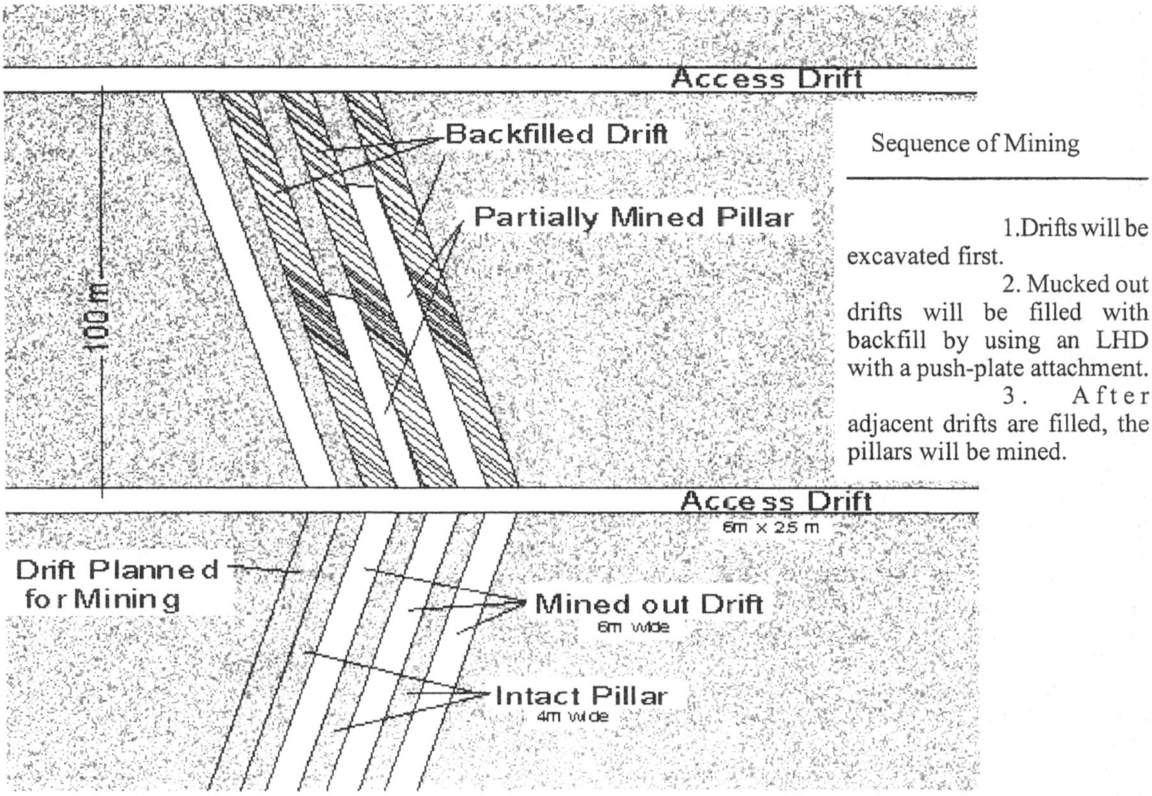

Figure 6. Schematic plan of the drift-and-rib-pillar stoping layout

and 0.88% Pb.

Kvaerner concludes that an appropriate, sustainable mining rate will be between 2 000 and 3 000 tons per day (tpd). At 3 000 tpd, the mine life would be just over 7.6 years. This mining rate is well within the capability of two roadheader units.

Kvaerner has also estimated a cut-off grade of 5.6% Zn at a price for zinc of \$ 0.50 per pound (\$ 1,100 per ton).

9 MINERAL PROCESSING

9.1 *Conventional Flotation*

From a casual first examination of drill core from the deposit, it is easy to conclude that recovery of zinc may be difficult because of the fine-grained nature of the mineralization. Indeed, test work done by Lakefield Research Limited in 1997 indicated that the best cleaner zinc concentrate contained 53.7% zinc with only a 62.9% zinc recovery. The conclusion was that pyrite in the zinc concentrate was responsible for this poor performance.

However, the same test work indicated that the rougher concentrate had a grade of 24.61% zinc with a 94% zinc recovery.

The recovery of lead is not thought to be economically viable and was not given serious consideration.

There may well be additional test work that could be done to improve the efficiency of producing an acceptable cleaner concentrate. However, since prospects for success did not seem obvious, other processing technologies were pursued.

9.2 *Leaching*

Kvaerner suggested that the project should investigate the proprietary zinc leach technology being developed by Drinkard Metalox, Inc. of Charlotte, North Carolina.

Test work was done by Drinkard on both the rougher concentrate and whole rock ore. While whole leaching may be feasible, it was decided to only consider the leaching of rougher concentrate in order to limit the size of the leach plant.

The test work conducted by Drinkard was very encouraging. The zinc recovery from the zinc concentrate was 98.4% while limiting the extraction of lead, iron, calcium and carbon. In addition, the process converted only about 50% of the sulphur to sulphate with the balance being present as elemental sulphur or unreacted sulphides. The bulk of the iron in the concentrate ends up in the leach residue either as hematite or unreacted pyrite.

The impurity removal stages were able to remove almost all of the impurities. Zinc oxide was achieved as the final product.

The Drinkard leach process takes place in open vessels without the addition of heat.

The advantages of producing zinc oxide instead of sphalerite concentrate from the mine are numerous:

1. Zinc oxide is 80% zinc instead of the 53% or so for conventional concentrate, thereby reducing the tonnage of mine product to be shipped for final refining;

2. Zinc oxide can be fed directly into the purification circuit at a zinc refinery and contains no iron or sulphur which are major disposal problems for refiners;

3. Expected market value of zinc oxide is considerably higher than for conventional zinc concentrates, although no market studies have been conducted;

4. There is the option of producing chemical grade zinc oxide at the mine site, which has a higher market value;

5. There is the option of producing electrolytic zinc metal at the mine site, if economics warrant in the future.

The complete processing circuit is summarized in Table 1.

Table 1. Overview of Process Circuit

Major Conventional Flotation Circuit Components
Crushed ore storage and reclaim
SAG and ball mill grinding and cyclone classification
Zinc sulphide rougher flotation circuit
Zinc concentrate dewatering and filtering
Tailings dewatering and filtering
Underground backfill preparation plant
Major Leach Circuit Components
Digestion circuit
Sulphate removal circuit
Gypsum dewatering and filtering
Iron precipitation circuit
Hematite dewatering and filtering
Zinc oxide circuit
Leach reagents regeneration circuit
Zinc oxide product storage

10 FACILITIES, INFRASTRUCTURE AND MANPOWER

The Citronen Fjord project is a remote operation and will need to provide all of its own infrastructure. The major components of the required infrastructure are listed in Table 2.

Table 2. Required Infrastructure

Major Infrastructure Components
Site roads
Air strip
Underground backfill plant
Tailings disposal pond for tailings not sent underground
Oxygen plant
Power plant electrical distribution system
Port facilities and ship loading
Product storage facility
Administration facility
Warehouse, assay lab, dry and shops
Permanent camp
Communication system
Fuel storage

There is no local source for hydroelectric generation potential identified. Therefore, it is assumed that all electric power generation will be diesel generated.

It is expected that operating crews will work 10 hour days (12 hour days for continuous operations), 7 days per week, on an in-out rotation basis of 2 to 1 (1 to 1 for continuous operations). Staffing levels are therefore about two-thirds higher than the level required on site at a particular time. The number of persons expected to be hired is about 200, split evenly between the mine and the processing operations.

11 SHIPPING

The Citronen Fjord location has never been accessed by sea. Large ice-breakers have been past Kap Morris Jessup, the northernmost point in Greenland, from the east, but not into Frederick E. Hyde Fjord. For the Citronen Fjord deposit to become a viable mine, it will have to be demonstrated that existing bulk carrier shipping can access Frederick E. Hyde Fjord on a regular basis over an acceptable annual operating period.

As an aid towards demonstrating that sea access to the project site is feasible, Greenland Resources A/S on contract from the Greenlandic Government's Bureau of Minerals and Petroleum subcontracted

Enfotec Technical Services Inc. (Enfotec) of Ottawa, Canada to complete a study to define the ice conditions and ship access to northeastern Greenland. Enfotec is particularly qualified to complete this study since it specializes in the analysis of ice conditions for marine shipping and operates an ice navigation service that supports shipping operations to all mines in northern Canada and Alaska where navigation through ice covered seas is required.

Enfotec's study included a detailed review of the historical distribution of ice conditions in the region based on satellite images including those obtained by the US Defence Department Meteorological Satellite Program. The review spanned the period 1978 to 1998.

The conclusion from the Enfotec study is that a DNV Polar 30 ice class ship could have a navigation season to the project site of between 130 and 60 (average 90) days per year. A ship of DNV Polar 20 class could have an average navigation season of about 60 days. A DNV Polar 20 class ship would have a 10% chance risk of not performing in any given year due to heavy ice conditions.

DNV Polar 30 class ships have yet to be built, although the technology exists. The MV Arctic which operates nearly year round into Deception Bay on the north coast of the Ungava Peninsula in Quebec, Canada, is an existing ship with most of the specifications of a DNV Polar 20 class vessel. The MV Arctic has performed yeoman service to the Polaris and Nanisivik mines in the Canadian high Arctic.

The results of the Enfotec study are encouraging and will serve as critical background information for a future test voyage into Frederick E. Hyde Fjord.

12 CAPITAL AND OPERATING COSTS

Kvaerner has estimated order-of-magnitude capital and operating costs for a mining operation as described above. The largest unknown cost factor is transportation. Naturally this is because no ship has ever been to Citronen Fjord. Even following a test voyage into the site, it will be difficult to estimate transportation costs since the service required is so unique and may require the building of a new vessel. Nevertheless, with this caveat in mind, transportation costs were factored using the Polaris mine location as a model.

It is expected that the mining preproduction will take about six months once the crew is on site and the plant is set up. It is estimated that the mining and processing facility could be ready for start-up and commissioning within 24 months.

Tables 3 and 4 summarize the capital and operating costs, respectively, grouped by major cost centre.

No attempt has been made to estimate the revenue side of a potential mine operation. The uncertainty of the transportation costs for getting the product to market and the uncertainty of the market value for bulk zinc oxide makes estimating revenues too unreliable to be useful at this time. This means that economic parameters like rate-of-return or payback period have not been estimated.

Table 3. Capital Costs

Major Cost Centre	$ millions*
Mining (equipment and material)	17.5
Milling (equipment and material)	14.1
Leach Plant (equipment and material)	8.7
Factored Commodities	26.4
Infrastructure	29.4
Power Plant	13.0
Indirects	38.8
EPCM	13.7
Allowance for Unexpected	32.3
Project Total	194.0

* $US 1998

Table 4. Operating Costs

Major Cost Centre	$ per ton*
Mining (drift-and-rib-pillar)	21.97
Milling	15.10
Leaching	11.34
Power Plant	3.56
Other	2.00
Total	53.97

* $US 1998

13 CONCLUSION

An underground mining and processing concept has been developed for the currently known sedex zinc deposits at Citronen Fjord. While the concepts may not be in use at any other operation in the proposed combination, none of the individual components rely on unproven technology.

The estimated capital costs of about $200 million and operating costs of about $60 per ton are well within the norm for deposits fo this type elsewhere. The biggest operational issue remaining is to demonstrate reliable shipping access - a test voyage will be required for this.

Figure 7. Looking north over Beach Zone into Citronen Fjord and north coast of Frederick E. Hyde Fjord

Sedex deposits are often grouped, as dramatically demonstrated at Red Dog in Alaska. The Esrum and West Gossan areas on the Citronen Fjord property offer attractive potential for increases in zinc resources. The mining concepts developed in this project should help to provide the encouragement to continue exploration in order to achieve a threshold economic resource at Citronen Fjord.

REFERENCES

Enfotec Technical Services Inc. 2000. *Final report, Definition of ice conditions and ship access to Northeastern Greenland.* Ottawa.

Glahder, C. & H.C. Langager 1993. *Reconnaissance in the Citronen Fjord area, North Greenland, August 1993.* Greenland Environ. Res. Inst. & Greenland Field Investigations, Copenhagen.

Glahder, C. & G. Asmund 1995. *Baseline study in the Citronen Fjord area, North Greenland 1994.* Greenland Environ. Res. Inst., Copenhagen.

Kvaerner Metals, Division of Kvaerner Canada Inc. 1998. *Platinova A/S, Peary Land project order-of-magnitude study.* Toronto.

Lakefield Research Limited 1997. *An investigation of the recovery of zinc from Citronen Fjord project samples submitted by Platinova A/S.* Lakefield, Ontario.

Platinova A/S 1997. *Annual Report to Shareholders.* Toronto.

Platinova A/S 1998. *Annual Report to Shareholders.* Toronto.

van der Stijl, F.W. & G.Z. Mosher 1998. The Citronen Fjord massive sulphide deposit, Peary Land, North Greenland: discovery, stratigraphy, mineralization and structural setting. *Geology of Greenland Surv. Bull. 179.*

Mining in the Arctic, Olsen, Lorentzen & Rendal (eds), © 2001 Swets & Zeitlinger, Lisse, ISBN 90 5809 177 5

Production to begin in 2001 at Svea North, Svalbard

R. Hermansen, Managing director
Store Norske, Longyearbyen, Svalbard

ABSTRACT: Store Norske has proven the Svea North deposit through 72 diamond core drillholes penetrating more than 200 m of glacier ice plus 200 m of rock. The mine holds 75 million tons of coal, of which 42 million will be extracted by longwall mining. At the outer boundaries the seam is split by a layer of rock. In this part the mine layout will be a modified longwall or room-and-pillar. Development work began in January 2000 by a three-entry system penetrating the side-moraine of the Høganes glacier. So far 10,000 metres of drifts have been achieved. Quality, structures, faults, water-conditions and methane gas have been monitored. No unforeseen conditions have been encountered. A total of NOK 600 million will be invested in the mine Svea North. A thorough environmental impact assessment is completed. The annual output will vary between 1.2 – 1.9 Mtpy, ensuring mining for the next 3 decades.

BACKGROUND

Store Norske Spitsbergen Kulkompani AS was established on 30 November 1916. Svalbard was known for its rich natural resources, and at the time the coal produced by the company was important for securing Norwegian energy supplies - particularly in northern Norway.

The company's objects clause: "The object of Store Norske Spitsbergen Kulkompani Aktieselskap is to exploit, through operations or otherwise, the company's properties and rights on Svalbard. Furthermore, the company may take part in and operate other associated activities".

This objects clause has meant that the company, throughout its life, has been mapping Svalbard's natural resources and claimed areas for exploitation. Areas claimed under the Mining Agreement for Svalbard, come with exploration and production rights to the minerals or deposits within the claim area.

Today, the company holds 316 claim areas, of which 142 are occupation claim areas. An occupation claim area is an entitlement which the company had before Norwegian sovereignty was established on Svalbard and which under the Svalbard Treaty and the Mining Agreement for Svalbard was converted to a claim area.

The claim areas secure Norwegian control of important natural resources such as coal, gold and petroleum. The company has made two unsuccessful attempts at drilling for oil. The gold deposit is rich and the company will explore further to establish whether the field is commercially viable. The company has demonstrated thick coal seams in the surveyed coal fields and further explorations will be carried out in the time ahead.

The field which is now due to be developed, Svea North, has been surveyed through surface mapping as well as exploration drifts. No unexpected circumstances or insurmountable problems have been found.

THE SVEA NORTH BUSINESS IDEA

The business idea behind the company's imminent new operation, is as follows: "To produce and sell coal in large enough quantities for the production to be profitable and to contribute to further fulfilment of the objects clause".

In January 1999, the Board of Directors submitted their recommendation of 29 January 1999. At this point, the company's activities in Mine 7 in Longyearbyen and Svea West were coming to an end, and it was important to explore the Central Field with a view to developing a new mine at Svea North.

The Board of Directors recommended that the Annual General Meeting fund the exploration work with government assistance. The Storting (Norwegian Parliament) gave their consent through the Revised National Budget for 1999. The exploration period was expected to last till the end of 2002.

The Storting committed themselves to a NOK 116 million government grant for funding the explorations and to cover the operating losses incurred during the exploration phase.

As the business idea is to produce and sell coal in quantities large enough for the coal operation to be profitable, the company has reviewed its cost structure and production process. Capacity studies have been carried out and bottle necks have been removed.

Svea has now been developed and is designed for long-term profitable production.

MAPPING THE FIELD

The field covers 20 km^2 and has been thoroughly explored through core drilling. Space between drill holes is 600-800 metres. Seam thickness is 2-5 metres, and the field holds a total of 75 million tonnes of coal.

The exploration drift has allowed the company to study the mining and operating conditions at Svea North. The rock pressure has been examined, as has the mechanical qualities of hanging walls and side-walls, waste rock, foldings, water and gas. These are all matters of significance to quality and operating plans. The operating plans ensure maximum safety. The explorations have found nothing unexpected nor any unsurpassable problem.

The main coal field is well suited for longwall stoping.

COAL RESERVES

Table 1 Core of deposit

			Mill. tonnes
Longwall stoping	(215 km at 250 m)	100%	26.5
Mine development	(300 km*4*25)	100%	3.0
Room and pillar,	pillars by the main drift	45%	1.1
Room and pillar,	South	60%	1.5
Total reserves for recovery			32.0

Table 2 The fringe zone

Svea East, longwall	70%	2.5
Lower deposit at the fringe of Svea North, longwall	70%	6.5
Total reserves for recovery		9.0

Explorations show that the company's resources may provide a basis for 20-30 years of operation with an annual production volume of 1-2 million tonnes.

THE PRODUCT

The Company recovers coal from the Svea field. The Svea coals are high-energy coals, well suited for mixing with other coal grades in large coal-fired power stations. Thanks to a low ash melting point, the coals are not suitable for use on their own in all kinds of power stations.

Average coals from Svea North have the following qualities (dry basis):

Energy	7500 Kcal
Ash	5%
Sulphur	0.6%

CLIMATE GASES

Due to global pollution levels, coal is a much-debated product. The production of coal releases methane, the transport and burning of coal release CO_2.

EMISSIONS OF METHANE AT THE PRODUCTION OF COAL

The company has been monitoring emissions of methane at the production of coal over a long period, as methane represents a safety hazard in the mines. Too high concentrations may cause explosions and major accidents involving loss of human life.

The methane levels measured at Longyearbyen have been between 0.5 m^3 and 0.9 m^3 per tonne. Measurements at Svea North have shown methane emissions of 0.8 m^3. This covers total emissions caused by the entire process, from stoping to burning as fuel.

Methane emission levels are normally between 8-60 m^3, most often approx. 20 m^3 per tonne. The emission levels are particularly low at Store Norske because the mines are situated at high altitudes and the rock is fractured to the extent that the methane has been aerated before the coal is recovered.

The fact that the methane emission levels are this low, gives the product an environmental advantage in the climate accounts.

TRANSPORT TO CONSUMERS

Store Norske's mines are situated a short distance from the export quay. This provides for lower on-shore transport costs and a low emission of CO_2 in connection with land-based transport.

Svalbard is situated a short distance from Europe. This means lower transport costs and lower transport-related CO_2 emission levels.

EMISSIONS AT BURNING

Coals from Svea North have a high energy content. This means that little energy is lost when they burn; which in turn means that emissions produced by burning the coal are lower per energy unit than for most other competing coal types when burnt in the same power station.

MARKET

Every year, approx. 4500 million tonnes of coal are produced world-wide. The largest producers are China, Australia, India and South Africa. In Europe, excluding the former Soviet Union, the production is 780 million tonnes. The largest producers in Europe are Germany, Poland, the Czech Republic and Greece.

In addition, approx. 150 million tonnes of coal are imported to Europe every year.

The company considers Europe to be its natural market.

World trade in coal represents approx. 500 million tonnes annually. The major exporters are Australia, South Africa, USA, Indonesia and Canada. The largest importers are Japan, Korea and the 15 EU countries. Germany, England and the Netherlands import 65 million tonnes every year, the Nordic countries import 21 million tonnes and Norway 2 million tonnes of coal and coke products.

In Norway, coal is largely used by the cement and metallurgical industries.

Coals produced at Svea are typical fuel coals, well suited for the cement industry and the generation of power.

This means that the company considers the power market in Europe its market place. As stated above, the annual take-up of coal in Europe is approx. 850 million tonnes. Store Norske is an extremely small player with its marketable production of 1.2 million tonnes.

Coal is sold through traders who control the market. Store Norske has therefore entered into a 10-year partnering agreement with SSM Coal bv, a Rotterdam-based company which will work as Store Norske's sales representative.

INFRASTRUCTURE

There have been mining activities in the Svea area since the beginning of the last century. In 1933 Store Norske bought the area from Swedish interests. The distance between Longyearbyen and Svea is relatively short, approx. 60 kilometers. There is no road between the two places, and in summer all transport to Svea is by boat or airplane. In the winter, a winter road is established for dozers and snowmobiles. It is less than 40 kilometres between Mine 7 and Svea North.

To ensure a predictable development of the place, an area plan has been drawn up with indications of land use. The plan accounts for expected developments over the exploration period. In the course of 1999 and 2000 a road was built between Kap Amsterdam, the export quay, and Svea North. The road up to Svea North has been laid on a glacier. Even if 2000, with its large volumes of precipitation, involved high maintenance costs, we see that the road will work well in the future.

Accommodation, water supply, baths, storage facilities and fuel plants have all been upgraded. Svea is designed to accommodate a staff of up to 230 people.

OPERATIONS

The exploration drifts which have already been made in the fringe zone will make up the mine's main drift system. The tunnels are designed with this in mind and are secured against collapsing for as long as the mine is in operation.

The development of the stope panels will be based on a two-drift system. This gives the greatest level of safety with respect to the rock tensions and the rock pressure occurring in the field. The length of each stope panel will be approx. 3 km.

A single CM will be used for mine development and production will be based on a single stope.

PRODUCTION CAPACITY

The company has acquired two CM sets. Each set consists of a Continuous Miner, 2-3 bolting rigs and two shuttle cars.

Each cut is 13 metres and each shift can make 2 cuts. This means a drive of 25 metres per shift. With one CM in operation per shift, our capacity will enable us to cut 25,000 metres per year.

Each new longwall panel, 250 metres wide and 3 kilometres long, requires 16,000-metre drift. With an annual production of 1.6 million tonnes of coal, we will need to drive 10,000 drift meters per year.

We will need to invest in longwall mining equipment. In order to secure project progression, the company has issued a letter of intent for the purchase of such production equipment. The equipment will have an immediate capacity of 2,500 tonnes per hour, is fully automatic and covers our production needs many times over. The economic life of this equipment will match the mine operation period. The price tag for the stoping equipment is NOK 170 million, inclusive of the lifting equipment which is required to move the equipment once a panel has been exhausted.

The capacity of today's conveyor plant is designed for production with CM. In order to transport the increased production out of the mine, the conveyor plant will need to be replaced. The investment costs are NOK 50 million.

In addition, it will be necessary to acquire a mining tractor, upgrade the mine entrance plant and the storage facilities. Total investments are estimated at approx. NOK 20 million.

Table 3 Staffing at Store Norske

	2001	2002	2003
Administration	34	17	17
Mine administration	21	15	12
Staffing infrastructure	12	17	15
Supervisors and operatives	152	61	38
Total, Svea North	233	110	82
Mine 7 in total	19	19	18
Store Norske in total	252	129	100
LNSN (loading and transport to mine entrance, and loading of ship)	5		
Kafe Busen (canteen and cleaning services)	15		
Air transport	5		
Total, directly related to coal mining activities	357		

THE WORKING ENVIRONMENT

The provisions for internal inspection laid down in the Norwegian Working Environment Act are complied with in full. The company has a separate Health, Environment and Safety Department headed by a Safety Manager who reports direct to the General Manager. Risk assessments are carried out for all work operations. Mine operation manuals are drawn up, and all applicable safety regulations are absolute. The company has also initiated a "best practice" project.

Work to reduce the number of accidents has been targeted, and the frequency of injures has been significantly reduced. The company's most important measure in this connection has been to introduce a safety-first-policy for all work operations.

HES rounds (health, environment and security) are organised annually, each year focusing on a different theme. The company's senior managers take part in this training schemes with fixed sessions, clearly conveying to all employees that safety concerns are anchored at the highest level. The company has its own mine rescue team and a well trained first aid team.

Mining in the Arctic, Olsen, Lorentzen & Rendal (eds), © 2001 Swets & Zeitlinger, Lisse, ISBN 90 5809 177 5

Challenges to exploration in Greenland's High Arctic plains and plateaus

S. M. Jensen & L. Thorning
Department of Economic Geology, Geological Survey of Denmark and Greenland (GEUS), Ministry of Environment and Energy, Thoravej 8, DK-2400 Copenhagen NV, Denmark

ABSTRACT: The most likely type of mineralisation in the extensive carbonate platform areas of North Greenland is epigenetic lead–zinc. Flat-lying and virtually undeformed Palaeozoic carbonate rocks cover the areas closest to the margin of the Inland Ice, forming rather featureless plains or plateaus separated by wide valleys. Quaternary glacial deposits are ubiquitous and tend to partially cover prospective ground and dilute or obscure the geochemical signature of underlying bedrock. This, and the somewhat anonymous petrophysical properties (especially conductivity) of lead and zinc minerals, to some degree limit the ability of regional geochemical and airborne geophysical surveys to detect mineralisation in these areas. Thorough fieldwork in prospective areas remains essential to localising targets for detailed geophysics and drilling. Remote sensing using hyperspectral image analysis, however, is a new technique that should have great potential as a regional exploration tool in these regions.

1 INTRODUCTION

Exploration in the High Arctic of North Greenland is met with many challenges; not only the remoteness and logistical impediments imposed by near-permanent sea-ice and the scarcity of airfields and fuelling facilities, but also factors determined by the geology and geomorphology of these regions.

Across North Greenland, nearly flat-lying and virtually undeformed strata of platform siliciclastics and carbonates occupy a 100–200 km wide belt adjacent to the margin of the Inland Ice. The most likely type of mineralisation to occur in this environment is epigenetic, carbonate-hosted lead–zinc.

Lead–zinc discoveries in northern Canada and North Greenland have more often that not been made visually by prospecting geologists, including most, if not all, of today's lead–zinc-producing mines in the Arctic.

2 GEOLOGICAL SETTING

North Greenland is dominated by sedimentary rocks of the Lower Palaeozoic Franklinian Basin that rest on reworked Archaean crystalline basement and Proterozoic sedimentary sequences. In Cambrian–Ordovician time the basin was characterised by a marked division into a wide siliciclastic and carbonate shelf to the south and a siliciclastic trough to the north (Figure 1). The Navarana Fjord Escarpment, traceable across most of North Greenland, constituted an abrupt shelf–trough transition. A reef belt with high carbonate build-ups lay on the outer carbonate platform, parallel to the escarpment.

Towards the end of the Ordovician, black, calcareous mud accumulated at a very low rate in the basin, interrupted by occasional debris flows (conglomeratic turbidites) derived from the carbonate platform. Some debris flows attained several hundred metres' thickness and covered thousands of square kilometres of the basin floor (Surlyk & Ineson 1992).

At Citronen Fjord in Peary Land (Figure 1) sedimentary exhalative sulphides formed in the three youngest mudstone intervals (separated by debris flows) of the Ordovician Amundsen Land Group. The depositional environment then changed drastically in the Early Silurian when sand turbidites derived from the emerging Caledonian orogen to the east began to rapidly fill the North Greenland parts of the Franklinian Basin.

In the Devonian – Early Carboniferous the siliciclastic slope and trough sediments were metamorphosed, folded and thrusted, forming the North Greenland part of the Innuitian Fold Belt, whereas the carbonate platform was left little deformed.

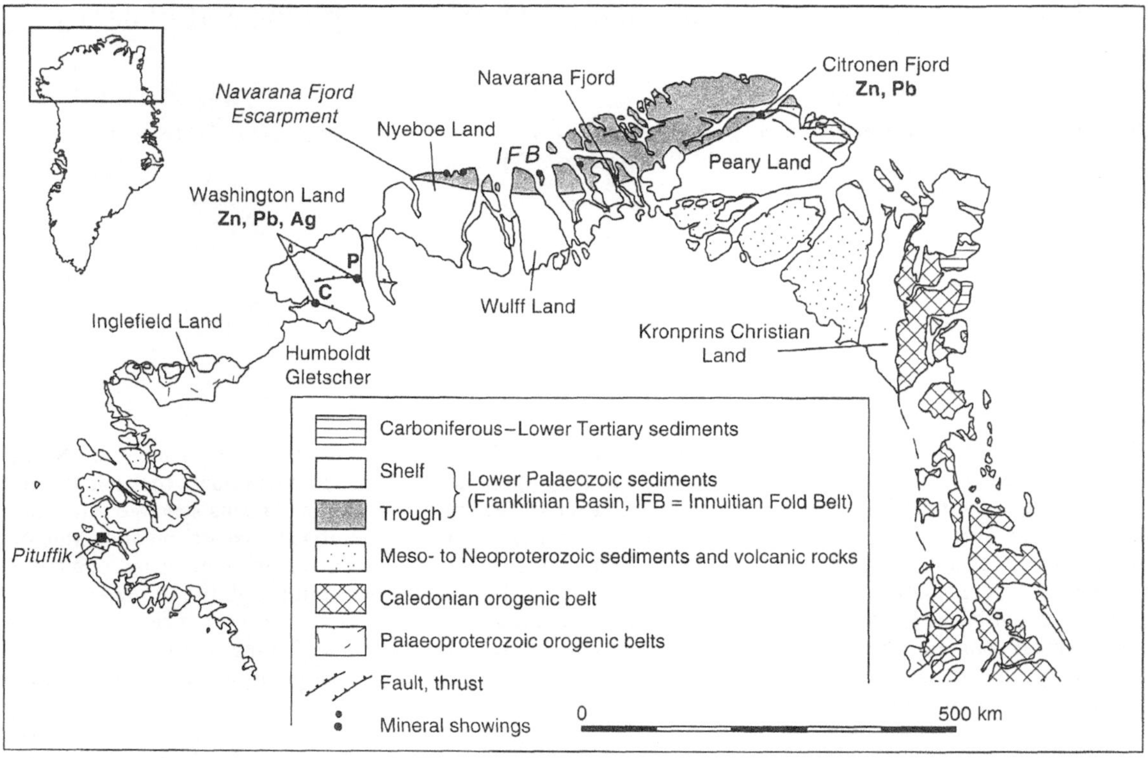

Figure 1. Simplified geological map of North Greenland. P = 'Petermann prospect'; C = 'Cass prospect'.

3 GEOMORPHOLOGY

Exploration and mapping experience from Kronprins Christian Land, Wulff Land and Washington Land (Figure 1) form the basis for the observations below:

The areas in question are categorised as plains or dissected plateaus in Dawes & Christie's (1991) geomorphic classification of the Arctic platform of Canada and Greenland. Central parts of Washington Land and Wulff Land are mesa-like tablelands (plateaus) separated by wide, flat-floored, glaciated valleys. In the southernmost parts of Washington Land and Kronprins Christian Land the plateaus give way to rolling plains.

The plateau surfaces are flat or gently rolling, often being made up of more resistive strata such as the Kap Coppinger Member (Cambrian) in Washington Land, a 10–25 m thick quartz-arenite unit with regional distribution (Figure 2). The plateaus commonly rise 400–800 m above the valley floors that generally lie 0–200 m above sea level.

North Greenland is an Arctic desert with cold winters and little precipitation (annual average temperature of -11.2°C and precipitation 124 mm at Pituffik). There are a few isolated valleys in areas of mainly siliciclastic rocks with relatively rich vegetation, but the carbonate-dominated areas are almost completely barren.

Figure 2. Landscape in central Washington Land with mesa-like tablelands separated by wide, glaciated valleys. The white and dark grey resistive units topping the mesa-like plateau are the Kap Coppinger Member (white; uppermost Cambrian), a 10–25 m thick quartz-arenite unit with regional distribution, and massive-weathering limestones of the Cape Clay Formation (dark grey; lowermost Ordovician). Vertical distance from valley floor to top of 'mesa' is approximately 800 m.

Surficial deposits in the form of till and frost-heaved boulders and regolith constitute variable, but ubiquitous cover on the plateaus, plains and in the valleys. Glacial erratic blocks and boulders of crystalline rocks are particularly common in central and southern Washington Land – although the nearest outcrops of crystalline rocks are in Inglefield Land to the south, separated from Washington Land by the 100-km wide Humboldt Gletscher (Figure 1). In fact, an assemblage of erratics collected in Washington Land has disclosed the existence of a hitherto unknown volcanic province under the Inland Ice somewhere to the southeast (Dawes et al. 2000).

The smaller mesa-like plateaus often have no drainage developed, except those that have isolated ice caps with seasonal runoff. Drainage of the plains tends to collect in few large rivers with high water flow that account for much of the overall sediment transport, whereas the smaller streams are generally shallow and gently flowing, carrying low sediment loads. Figure 3 shows a characteristic drainage system on the carbonate plains of Kronprins Christian Land.

Figure 3. Drainage system on the carbonate plain in Kronprins Christian Land, photographed approximately 25 km from the Inland Ice. The fine (< 0.1 mm) fraction of sediment samples in this region can have SiO_2 contents up to 85%, irrespective of underlying lithology. The high proportion of exotic, fine-grained quartz in the surficial deposits is attributed to Quaternary glacial and recent aeolian transport.

4 MINERAL OCCURRENCES

The most significant and voluminous mineralisation of the Franklinian Basin of North Greenland detected to date is the Citronen Fjord massive sulphide deposit (Figure 1) with a resource of some 20 Mt with 7% zinc and 1% lead (van der Stijl & Mosher 1998, Andersen, this volume). Massive and finely laminated sulphides are hosted in calcareous mudstones of the deep-sea trough, deposited just below the steep carbonate platform margin. The Citronen Fjord deposit is a particularly well-preserved example of sedimentary-exhalative (SEDEX) mineralisation (Kragh et al. 1997, van der Stijl & Mosher 1998, Kragh 2000).

Mineralisations set in an outer shelf-and-slope environment occur in the Navarana Fjord anticline and westwards along the coast to northern Nyeboe Land (Figure 1). In the Navarana Fjord anticline, Ordovician carbonate debris flows carry minor, disseminated zinc mineralisation in the matrix (von Guttenberg & van der Stijl 1993, Stemmerik et al. 1996). A number of lead–zinc occurrences lie in folded Ordovician outer shelf-and-slope rocks along the north coast between Navarana Fjord and Nyeboe Land (von Guttenberg & van der Stijl 1993).

Cambrian platform carbonates at Navarana Fjord host a 5–7 m wide, brecciated calcite–baryte vein, with sphalerite, smithsonite and galena (Jakobsen & Steenfelt 1985). The vein has been traced intermittently for a few kilometres east–west.

Showings in the undeformed Ordovician carbonate platform of North Greenland are few in number. Two mineralised areas in Washington Land, being the most significant prospects in this setting to date, are described briefly below:

4.1 *'Petermann prospect'*

A new zinc–lead–silver occurrence in eastern Washington Land ('Petermann prospect'), was discovered in 1997 (Jensen & Schønwandt 1998). The mineralisation is clearly epigenetic, and associated with extensive dolomitisation. The showings occur over a 10-km segment of a major east–west fault system that can be traced for about 50 km in east-central Washington Land (Figure 1). The showings were noted as rusty weathering patches with frost-heaved sulphide-mineralised boulders and regolith (Figure 4). True outcrops of mineralisation are rare, and are found primarily in river gorges (Figure 5).

Some of the characteristics of the Polaris deposit in Nunavut were cited as possible analogies to the mineralisation at 'Petermann prospect' (Jensen 1998, Jensen & Schønwandt 1998): overall stratigraphic setting; fault-controlled mineralisation; pervasive dolomitisation of host rocks; evaporitic units in sedimentary succession being a possible source of sulphate; hydrocarbons (now bitumen) may have thermochemically reduced the sulphate to sulphide.

A drilling programme comprising ten holes (925 m of core in total) over 5 km of the mineralised trend conducted in 1999 by Platinova A/S and Rio Tinto Mining and Exploration Ltd., however, failed to intersect significant sulphide mineralisation, except for two intersections in one hole of massive pyrite with a combined thickness of 16.8 m (Platinova A/S 2000).

Figure 4. Looking east along the 'Petermann prospect' in eastern Washington Land. The geologists are standing on rusty weathering, frost-heaved and strongly dolomitised rock/regolith with the zinc minerals smithsonite and hydrozincite, galena and pyrite. Such patches of mineralisation occur intermittently on the relatively flat area between the ice-covered lake and the mountain side, and isolated mineralised boulders have been found along the strike of a regional fault as far away as about 8 km to the east and about 15 km to the west of this locality.

Figure 5. Outcrop of Kap Coppinger Member quartz-arenite beds in the 'Petermann prospect' area, Washington Land. Approx. 15 m of till covers the bedrock, which is only exposed in the river gorge. Height of outcrop approx. 30 m.

4.2 *'Cass prospect'*

During Platinova's 1999 exploration work, a new area of mineralisation ('Cass prospect') was found some 50 km south-west of the 'Petermann prospect' (Figure 1). Lead and zinc occur with baryte in faulted and dolomitised limestone. Eight showings have been found over a 4–5 km segment of a regional fault. The dolomitisation is distinctly reddish, suggesting ferroan dolomite. The presence of baryte and ferroan dolomite distinguishes this mineralisation from that of the 'Petermann prospect' which is characterised by white or light buff dolomite. One drill hole has been completed on the 'Cass prospect' with an intersection of 1.2 m with 8.4% zinc, 0.04% lead and 94 g/t silver (Platinova A/S 2000).

5 CHALLENGES TO EXPLORATION

Some of the challenges to regionally applied exploration methods posed by the geology and geomorphology of the carbonate plains and plateaus of North Greenland are discussed below:

5.1 *Geophysics*

Lead-zinc mineralisation commonly does not have petrophysical properties that make it a distinctive feature in airborne geophysical surveys. Most importantly, the conductivity, even of relatively pyrite-rich deposits, can be poor.

Gravimetric surveys are potentially useful for detecting carbonate-hosted lead–zinc mineralisation and its common envelope of dolomitisation, but may not be able to distinguish this from barren, pervasive dolomitisation. Washington Land has many areas of intense dolomitisation that, on closer inspection, show no signs of sulphide mineralisation.

If mineralisation is related to structures such as faults, an airborne geophysical survey may detect the structure and give an indication of its 'magnitude'. A government-financed airborne electromagnetic and magnetic survey in 1998 clearly outlined a major structure coinciding with the mineralisation-controlling fault of the 'Cass prospect'. The regional mineralisation-controlling east–west fault system of the 'Petermann prospect', on the other hand, was not readily identified by the airborne geophysical survey.

Four clusters (each a few kilometres in diameter) of weak electromagnetic anomalies identified for follow-up (Rasmussen 1999) were checked by field parties in 1999. No mineralisation was detected in any of the areas, but this does not constitute proof against the possibility of mineralisation: bedrock geology was in every instance covered by surficial deposits. Many of the anomalies may be attributable to clayey material in frost boils and outwash channels from local ice caps, or to evaporites in parts of the stratigraphy (Figure 6; Rasmussen 1999). In the 'Petermann prospect' area a strong electromagnetic response from the nearby lake (Figure 4) complicates the interpretation of the anomaly pattern. Two intriguing questions remain unanswered: why do the electromagnetic anomalies occur in clusters; and is there mineralisation below the till cover?

Figure 6. Ground check of an electromagnetic anomaly in southwestern Washington Land. The relatively weak electromagnetic anomaly appears to be related to frost boils developed in a broad outwash channel on surficial deposits of glacial origin. Soil samples from the frost boils do not reveal the source of the anomalies. Is there mineralisation below the surficial deposits?

5.2 *Geochemistry*

Glacial or aeolian deposits can blanket out, dilute or obscure the geochemical signature of underlying bedrock. For example, the fine fraction (< 0.1 mm) of stream sediment samples collected in a ~ 22,000 km^2 area of mainly Ordovician limestones and Proterozoic sandstones and volcanics in North-East Greenland had up to 85% SiO_2 in a 40-km wide zone parallel to the margin of the Inland Ice, irrespective of underlying bedrock lithology (Jensen et al. 1994, 1995). Wind-eroded carbonate outcrops, faceted pebbles on rocky pavements and dunes in some of the valleys attest to frequent sand and dust storms in these areas. The effect of the high contents of exotic fine-grained quartz is to 'dilute' the true geochemical signature of the underlying lithologies.

Another example is from Washington Land: Although there are no outcrops of crystalline basement in Washington Land, crystalline erratic boulders occur in streams and on the plains is most parts of this isolated landmass. The heavy mineral fraction of stream sediments, again collected in limestone-dominated areas, consists mainly of garnet and magnetite, minerals that do not occur in significant amounts in exposed bedrock lithologies in Washington Land (S.M. Jensen, unpublished data).

Soil (or finely divided regolith) sampling on the 'Petermann prospect' has produced highly anomalous results. A sampling programme conducted on the most strongly dolomitised area returned astonishing zinc values of up to approximately 40% (Platinova A/S 1999). Likewise, a sample from soliflucted soil at the main showing of the 'Cass prospect' returned anomalous zinc values, whereas the 'normal' fine fraction of stream sediment from a stream running through the mineralised area did not detect the mineralisation (S.M. Jensen, unpublished data). Platinova reports mineralised float collected along this river up to 2 km downstream from its intersection with the mineralised trend (Platinova A/S 2000).

Sampling of soil (or regolith) thus appears to be better suited than the fine fraction (< 0.1 mm) of stream sediment for detecting and mapping the extent of lead–zinc mineralisation in the carbonate-dominated plains. The area representation of a soil sample is of course much smaller than that of a stream sediment sample. In practice, systematic soil sampling, especially in the High Arctic regions discussed here, is only economically feasible on localised targets of moderate size.

5.3 *Remote sensing*

Remote sensing exploration techniques have been applied to a limited extent in North Greenland, but should have great potential in the large vegetation-free plains and plateaus.

Hyperspectral image analysis techniques seem to be particularly promising (see Aastrup et al., this volume). For example, it is conceivable that dolomitisation with associated sulphide mineralisation can be distinguished from barren dolomitisation; and it may be possible to detect weak smithsonite or hydrozincite haloes in surficial material around mineralisations.

6 CONCLUSIONS

Exploration in the carbonate plains of North Greenland is complicated by a combination of factors:

The flat-lying strata of undeformed carbonate rocks form plains or mesa-like plateaus separated by wide valleys, susceptible to extensive glacial cover.

Lead-zinc mineralisation often does not give a strong electromagnetic response. Local gravimetric surveys have proven useful both at Citronen Fjord and in Washington Land, but may sometimes detect dolomitisation rather than lead-zinc sulphides; as noted earlier, there are many occurrences of unmineralised dolomitisation in an area such as Washington Land. Field checks of electromagnetic anomalies in Washington Land have not produced conclusive results; clusters of anomalies have been found to be in areas with rather thick surficial cover where they may or may not overlie mineralised bedrock.

Variable surficial cover can blanket out or dilute geochemical anomalies from underlying bedrock. The fine fraction (< 0.1 mm) of stream sediment can have a large component of glacial or aeolian exotic material, especially SiO_2. Soil sampling appears to be better able to detect mineralisation, but is only feasible on localised targets.

Hyperspectral image analysis, with its ability to detect subtle mineralogical differences and thin coatings of alteration minerals and thus to map out dispersion patterns, should be an ideal exploration tool in this type of terrain. The first use of hyperspectral image analysis in Greenland is described elsewhere in this volume by Aastrup et al.

REFERENCES

Dawes, P.R. & Christie, R.L. 1991: Geomorphic regions. In: Trettin, H.P. (ed.): *Geology of the Innuitian orogen and Arctic platform of Canada and Greenland.* Geology of Canada 3, 29-56. Ottawa: Geological Survey of Canada (also *The Geology of North America* E, Geological Society of America).

Dawes, P.R., Thomassen, B. & Andersson, T.I.H. 2000: A new volcanic province: evidence from glacial erratics in western North Greenland. *Geology of Greenland Survey Bulletin* 186, 35-41.

Jakobsen, U.H. & Steenfelt, A. 1985: Zinc mineralisation at Navarana Fjord, central North Greenland. *Rapport Grønlands Geologiske Undersøgelse* 126, 105-109.

Jensen, S.M. 1998: Carbonate-hosted Zn-Pb-Ag mineralisation min Washington Land, western North Greenland. *Geology of Greenland Survey Bulletin* 180, 67-72.

Jensen, S.M. & Schønwandt, H.K. 1998: A new carbonate-hosted Zn-Pb-Ag occurrence in Washington Land, western North Greenland. *Danmarks og Grønlands Geologiske Under-søgelse Rapport* 1998/3, 31 pp.

Jensen, S.M., Steenfelt, A. & Dam, E. 1994: Reconnaissance geochemical mapping of Lambert Land and southern Kronprins Christian Land (79° to 80°55'N, 18°50' to 25°W), eastern North Greenland. *Open File Series Grønlands Geologiske Undersøgelse* 94/19, 19 pp.

Jensen, S.M., Steenfelt, A., Dam, E. & Kalsbeek, F. 1995: Reconnaissance geochemical mapping of Kronprins Christian Land and Mylius Erichsen Land (79°40' to 81°20'N, 14° to 28°W), eastern North Greenland. *Open File Series Grønlands Geologiske Undersøgelse* 95/7, 20.

Kragh, K. 2000: The Citronen Fjord sedimentary exhalative sulphide deposit, North Greenland. A study of primary and di-agenetic textures, oxygen and carbon isotopes, and fluid inclu-sions, 321 pp. Unpublished Ph.D. thesis, University of Aarhus, Aarhus, Denmark.

Kragh, K., Jensen, S.M. & Fougt, H. 1997: Ore geological studies of the Citronen Fjord zinc deposit, North Greenland: project 'Resources of the sedimentary basins of North and East Greenland'. *Geology of Greenland Survey Bulletin* 176, 44-49.

Platinova A/S 1999: *Annual report 1998*, 20 pp.

Platinova A/S 2000: *Annual report 1999*, 20 pp.

Rasmussen, T.M. 1999: Airborne electromagnetic and magnetic survey in Washington Land and Daugaard-Jensen Land, western North Greenland. *Danmarks og Grønlands Geologiske Undersøgelse Rapport* 1999/10, 19 pp.

Stemmerik, L., Jensen, S.M., Korstgård, J., Schønwandt, H.K., Surlyk, F., Clausen, O.R., Fougt, H., Kragh, K., Langdahl, B.R. & Therkelsen, J. 1996: Resources of the sedimentary basins of North and East Greenland – an integrated petroleum and ore geological research project. *Bulletin Grønlands Geologiske Undersøgelse* 172, 32-36.

Surlyk, F. & Ineson, J.R. 1992: Carbonate gravity flow deposition along a platform margin scarp (Silurian, North Greenland). *Journal of Sedimentary Petrology* 62, 400-410.

van der Stijl, F.W. & Mosher, G.Z. 1998: The Citronen Fjord massive sulphide deposit, Peary Land, North Greenland: dis-covery, stratigraphy, mineralization and structural setting. *Geology of Greenland Survey Bulletin* 179, 40.

von Guttenberg, R. & van der Stijl, F.W. 1993: North Greenland project 1992, 65 pp. Unpublished report, Strathcona Mineral Services Limited and Platinova A/S, Toronto.

Mining in the Arctic, Olsen, Lorentzen & Rendal (eds), © 2001 Swets & Zeitlinger, Lisse, ISBN 90 5809 177 5

Geologic Setting of the Meadowbank Iron Formation-Hosted Gold Deposits

R. L. Sherlock
Canada-Nunavut Geoscience Office, P.O. Box 2319, Iqaluit, Nunavut, XOA OHO

R. B. Alexander, R. March & J. Kellner
Cumberland Resources Ltd., 906-595 Howe St., Vancouver, British Columbia, V6C 2T5

W. A. Barclay
W. A. Barclay Exploration Services Ltd., 23 Grenadier Rd., Toronto, Ontario, M6R 1R1

ABSTRACT: The Meadowbank gold deposits are located in the Archean Woodburn Lake group of the western Churchill Province. The stratigraphy, that hosts the deposits, consists of felsic to intermediate volcaniclastic rocks and interbedded iron formation. The stratigraphic sequence is polydeformed with four phases of deformation recognized regionally; two of which have substantially modified the geometry of the strata. The initial phase of deformation was a progressive transposition of the stratigraphy with tight to isoclinal folding which culminated in low angle high strain zones. This progressive deformation event appears to have controlled the distribution of gold mineralization. Overprinting, NE-SW shallow plunging folds further modified the geometry of the mineralized body.

1 INTRODUCTION

The Meadowbank gold deposits are polydeformed auriferous iron formations, within the Woodburn Lake group, of the western Churchill Province (Figure 1). The deposits are located about 70 km north of Baker Lake and are 100% owned by Cumberland Resources Ltd. who are currently conducting a prefeasibility study. Cumberland has identified a classified resource (Table One) of about 2 million ounces of which 962,400 are considered to be a proven and probable open pit reserve (Cumberland Resource Ltd., Press Releases March 29, 2000; April 27, 2000).

2 REGIONAL GEOLOGY

The geology of northwest Hudson Bay consists of Archean and Proterozoic supracrustal sequences and plutonic rocks (Figure 1). The Archean supracrustal sequences (e.g. Woodburn Lake and Prince Albert groups) are intruded by (*ca.* 2.60 Ga) orogenic granites, which are in turn unconformably overlain by pre-Hudsonian Proterozoic basins (e.g. Amer and Hurwitz groups). Post-Hudsonian basins (e.g. Baker Lake and Thelon Basins) unconformably overlie both sequences forming an overlap assemblage. The Woodburn Lake group is Archean in age, but has been overprinted by the Paleoproterozoic Hudsonian orogeny (Henderson et al., 1991). The extent of Proterozoic reworking and the implications for gold mineralization in the district is uncertain.

The Woodburn Lake group (Figure 2) includes quartzite and oligomictic conglomerate, ultramafic to mafic volcanic rocks, felsic to intermediate volcanic rocks and derived sediments with interbedded iron formation (Henderson et al., 1991; Zaleski et al., 1997b). Several phases of deformation have affected the stratigraphy with four events recognized regionally (Pehrsson et al., 2000). The supracrustal sequences have been intruded by a series of felsic rocks, some of which preserve D_1 and D_2 fabrics.

Woodburn Lake group felsic volcanic rocks, have been dated at 2.74-2.71 Ga and 2.63 Ga (Zaleski et al., 2000; Davis and Zaleski, 1998) suggesting two volcanic sequences (Davis and Zaleski, 1998). Felsic intrusives, east and west of the Meadowbank deposits give ages of 2612 ± 4 Ma (Figure 2; Roddick et al., 1992). Porphyroblastic biotite from the Third Portage area gave a K-Ar age of 1791 ± 32 Ma (Armitage et al., 1996).

Table One
Classified Resources Meadowbank Deposits

Class	Tonnes	Grade (g/t)	Ounces
Measured	1,191,000	6.58	252,100
Indicated	6,584,000	5.65	1,196,200
Inferred	3,467,000	5.60	623,800
Total	11,242,000	5.73	2,072,100

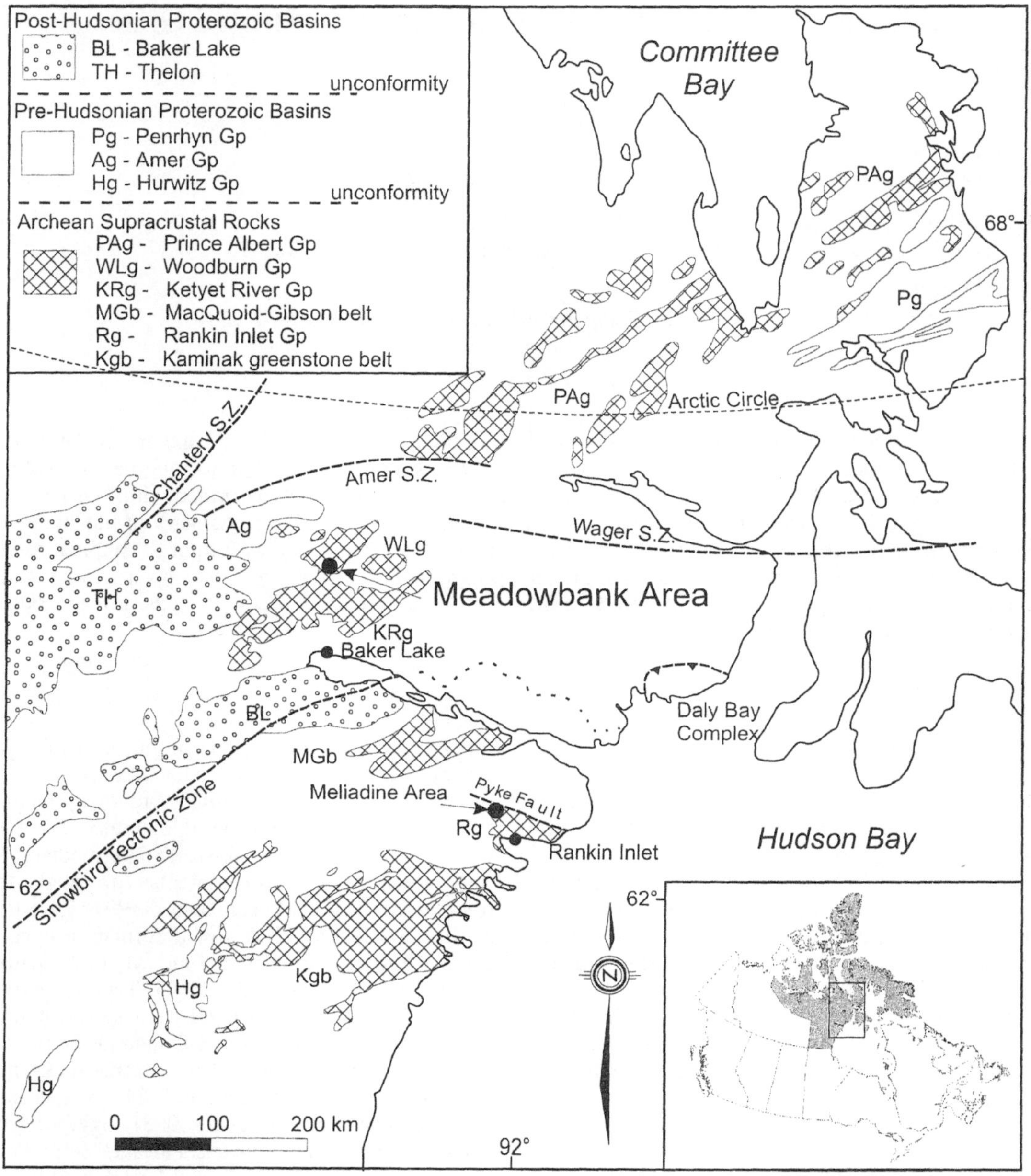

Figure 1. Regional geology of the western Churchill province showing the distribution of supracrustal rocks. The Meadowbank deposit is located within the Woodburn Lake group, about 70 km north of Baker Lake.

3 GEOLOGY OF THE MEADOWBANK AREA

The Meadowbank area (Sherlock et al., 2001; Figure 3) has been the focus of exploration activities since its discovery by Asamera Minerals Inc. in 1987 (Armitage et al., 1996). Most of the work in recent times (1995-1999) has focused on definition of the Third Portage, North Portage, Goose Island and the Bay Zone deposits, collectively referred to as the Meadowbank deposits. Exploration has been focussed on defining the geometry of the mineralized bodies *via* closely spaced diamond drilling and trenching. The following describes the lithologies observed on the property and the sequence of structural events that have deformed them.

3.1 *Felsic-Intermediate Volcanic Rocks (IV)*

The felsic to intermediate volcanic rock package is the dominant lithology recognized in the Meadowbank area. This unit may be subdivided into several subtypes:

Figure 2. Simplified geology map of the Woodburn Lake group, Meadowbank area (Zaleski et al., 1997a; 1999b).

- Massive felsic flows to subvolcanic intrusives (Figure 3) are recognized to the east, in the structural footwall of the mineralized intervals at Third Portage. This unit is quartz and plagioclase feldspar phyric with a groundmass of fine-grained quartz and biotite. Small quartz phenocrysts are abundant in this unit; commonly blue colored.
- Fine bedded volcanics (IV_b ; Figure 3) crop out immediately east of the massive felsic flows-intrusives. Bedforms are commonly 1-5 cm and rarely show unambiguous internal sedimentary structures such as graded bedding. Beds are composed mainly of quartz and feldspar grains, which display a granular clastic texture, along with an assemblage of epidote-biotite-chlorite-muscovite. Locally blue quartz phenocrysts occur similar to the massive units.
- Medium bedded volcanics (IV_b ; Figure 3) are commonly interbedded with the iron formations of the main mineralized intervals. These units are often relatively fine-grained (sand sized) and form beds 20 cm to 3 meters thick. Primary sedimentary features are rare. The beds are composed dominantly of quartz and feldspar grains with variable amounts of biotite-epidote-muscovite-chlorite. The beds are well sorted and homogenous in composition but are not graded, suggesting mass flow emplacement from a homogenous source. Alternatively these may represent agglomerated volcaniclastic beds. Locally blue quartz phenocrysts occur similar to the massive units.
- Finely foliated volcanics (IV_f; Figure 3) crop out to the west of the Third Portage zone, within the structural hanging wall of the mineralization. Bed forms are rare and the lithology is dominated by muscovite-chlorite-epidote with lesser quartz-feldspar grains. There exists a spectrum of rocks between the finely foliated, fine and medium bedded volcanic rocks, depending on their grain size, mineralogy and the development of bed forms. The dominant lithologies are shown in Figure 3.

3.2 *Iron Formation*

Intervals of iron formation comprising an assemblage of magnetite, quartz and amphibole are interlayered with epiclastic sediments of the felsic-intermediate volcaniclastic rocks. Several iron formations have been identified, including the East BIF, Central and West BIF (Figure 2). Of these only the Central BIF is auriferous. In areas of low strain, beds within the iron formation are typically 0.2 to 10 cm thick and consist of alternating mono-mineralic layers of magnetite, quartz ± amphibole. To the south, near Goose Island, garnet and biotite are developed in the iron formation, suggesting higher metamorphic grades.

3.3 *Quartzite*

Massive to bedded and foliated quartzite are exposed (Qtz; Figure 3) in the western portion of the Meadowbank area in the structural hanging wall of the deposit stratigraphy. The bedded material has centimeter-scale bed forms defined by selvages of muscovite and epidote. The foliated quartzite lacks bedforms and has correspondingly more muscovite which define the foliations.

At the base of the quartzite, oligomictic conglomerates have been intersected in drill core. These units range from less than 1 meter to over 10 meters in thickness. Local grading bedding provides younging directions. The conglomerates are dominated by quartzite fragments but also contain dark grey siliceous fragments as well as flattened fragments likely of mafic-ultramafic origin. Fuchsite-rich fragments are relatively common.

3.4 *Ultramafic Rocks*

The ultramafic rocks comprise a massive sequence of talc-amphibolite-chlorite with textural variations in the size and color of the ampiboles. This unit does not crop out in the immediate Meadowbank area but is commonly intersected in drill core. Only rarely are primary features such as spinifex textures preserved. Massively crystalline amphibole-rich rocks, with minor talc, are common near the footwall contact of the ultramafic package. Locally, within the ultramafic sequence, mafic volcanic intervals have been identified.

3.5 *Mafic Tuffs*

Intervals of mafic tuff, interbedded with the iron formation, are recognized locally in the felsic to intermediate volcanic package. These are narrow, generally less than 20 cm thick, and composed of chlorite and coarse biotite porpyroblasts.

3.6 *Intrusive Rocks*

The supracrustal rocks at Meadowbank are bracketed to the east and west by large granitic intrusive bodies (Figure 2). These intrusives locally preserve the dominant S_1 fabric described below. Locally, quartz and feldspar phyric dykes are intersected in drill core between the Third and North Portage deposits. These bodies are discontinuous and locally contain Cu-Au mineralization. Narrow undeformed lamprophyre dykes are also recognized in drill core.

3.7 *Metamorphism*

The metamorphic assemblages at Meadowbank are characterized by chlorite-epidote-muscovite-

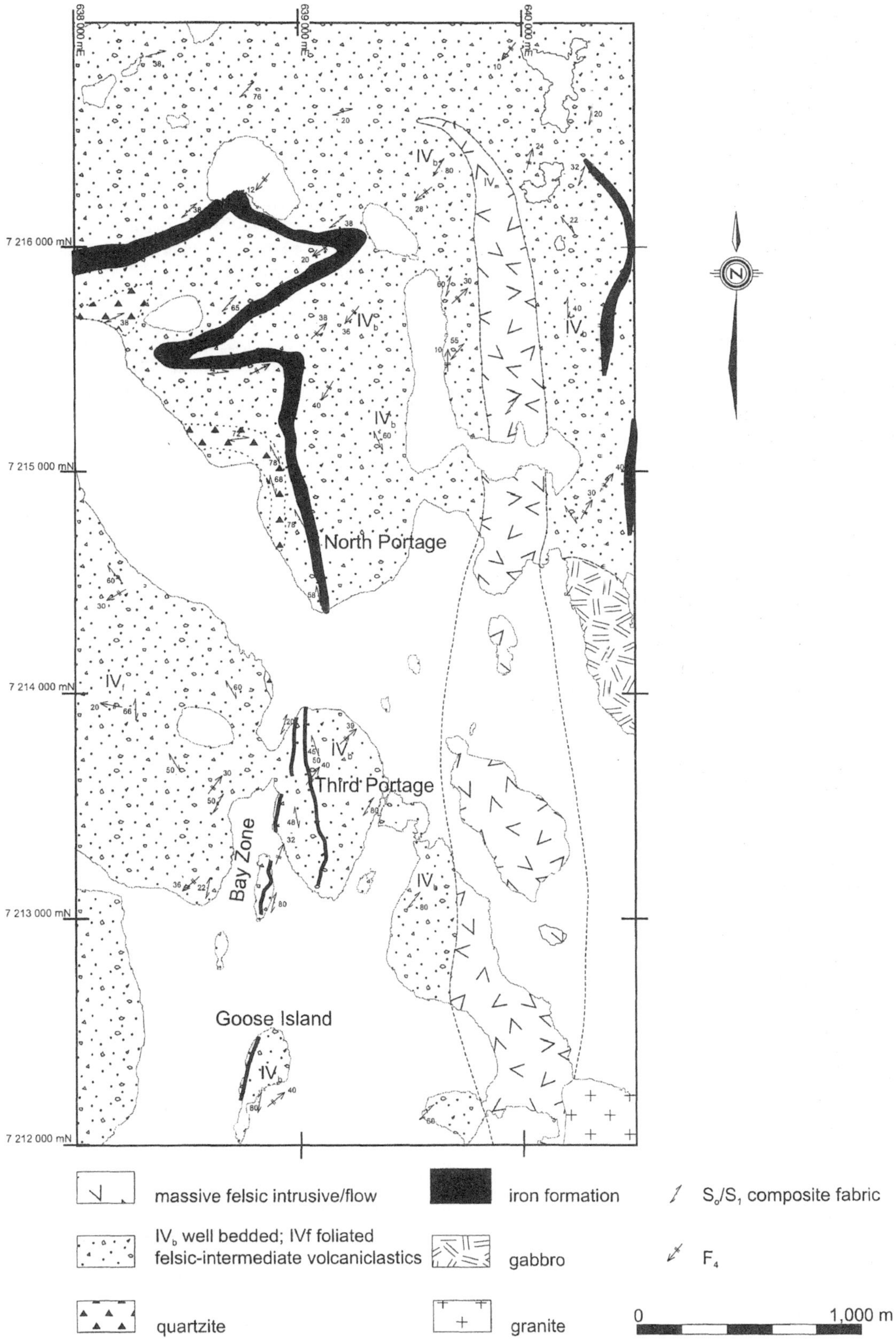

Figure 3. Geology of the immediate Meadowbank area (Sherlock et al., 2001).

biotite-calcite in the volcanosedimentary strata; interpreted to reflect regional greenschist facies metamorphic conditions. The iron formations are characterized by an assemblage of quartz-amphibole-magnetite ± sulphides. To the south, at Goose Island, the iron formations are characterized by a quartz-amphibole-biotite-garnet mineral assemblage. This suggests a transition from greenschist at the Third Portage to lower amphibolite metamorphic assemblages at Goose Island with the garnet isograd transecting Third Portage and Goose Island deposits.

3.8 *Structure*

The volcanosedimentary sequence at the Meadowbank area is polydeformed with four events recognized regionally (Pehrsson et al., 2000) two of which have significantly effected the geometry of the mineralized bodies. To clarify terminology we will adopt the deformation nomenclature of Pehrsson et al. (2000).

3.8.1 D_1-D_2

D_1 consists of a penetrative composite transposition foliation (S_0/S_1) and shallow plunging often rootless isoclinal folds that are particularly well developed in the iron formation. D_2 consists of a spaced to penetrative fabric (S_2) and tight to isoclinal, generally intrafolial, folds (F_2) which are shallowly N-S plunging and NE-E verging in the Third Portage area and NW-N verging in the North Portage area.

The transposed composite S_0/S_1 fabric is the principal fabric observed. The degree of S_1 development indicates that compositional layering preserved throughout the area should not be considered primary. Poles to S_0/S_1 (Figure 4a, b) form a weak girdle suggesting folding around a moderately plunging NE trending axis, consistent with F_4 described below. Coarse-grained amphibole rosettes overprint D_1 related fabrics.

D_1–D_2 deformation is considered here as a progressive event with refolding or buckling of the transposed S_0/S_1 fabric, generating tight to isoclinal folds (F_2) which exhibit moderately west dipping enveloping surfaces in the Third Portage area. The culmination of D_1–D_2 deformation involved the development of penetrative S_2 fabrics and locally low angle high strain zones axial planar to F_2 (Figure 4c, d). These high strain zones, often exploit lithologic contacts between ultramafic and volcanosedimentary rocks where rheologic contrasts have focused strain gradients into the volcanosedimentary rocks.

These strong S_2 fabrics crenulate amphiboles, suggesting either several generations of metamorphic amphiboles or that D_2 was post-peak metamorphism.

3.8.2 D_4

Folded D_1 – D_2 fabrics are typically open to close, angular to subrounded and generally "S" shaped in the Meadowbank area. Axial surfaces are upright, NE-SW trending and doubly plunging (Figure 5a, b, c). The geometry of F_4 is locally controlled by the orientation of the earlier fabrics (Pehrsson et al. 2000). Fold interference patterns are common resulting from the superposition of upright F_4 folds on tight and inclined F_2 folds. A spaced to penetrative axial planar cleavage (S_4) is locally well developed and overprints earlier fabrics as well as crenulating amphibole rosettes. D_4 fabrics fold and deform mineralization.

There are weakly developed, generally open warps of earlier fabrics in the Meadowbank area which may represent an additional phase of deformation, possibly D_3 of Pehrsson et al. (2000). Locally there is a weak NW-N axial planar spaced cleavage which crenulates earlier fabrics and also deforms amphibole rosettes. The effect of this event on the overall geometry of the mineralized bodies is minor.

4 GOLD MINERALIZATION

There are four discrete areas where gold occurs in sufficient concentrations to have justified a resource figure in the Meadowbank area. These include from north to south the North Portage, Third Portage, Bay Zone and Goose Island deposits (Figure 3). Of the four areas, the Third Portage deposit is volumetrically the largest and closest to the surface. The mineralization at Meadowbank should be best considered as several auriferous iron formations within a broad stratigraphic package. Each of the zones show many similarities and are located within the same broad structural regime.

The distribution of gold in the Third Portage deposit is well defined by closely spaced drilling (~25 m) and surface trenching (Figure 6), and is similar in style and geologic setting to the adjacent North Portage, Bay Zone and Goose Island deposits. Mineralization, at the Third Portage deposit, is best developed at the lithologic contact between an ultramafic body and the volcanosedimentary package. The ultramafic body may represent the core to an east verging antiform or alternatively may be a boudin within the transposed/flattened stratigraphy (Figure 6). Mineralization is almost entirely restricted to the structural footwall or the hinge area of the ultramafic body; occurring only sporadically and weakly in the structural hanging wall of the ultramafic unit. The hanging wall is low-strain with planar lithologic contacts and S_0/S_1 fabrics, whereas the footwall is generally higher strain with contorted F_2 folds and strong S_2 fabrics.

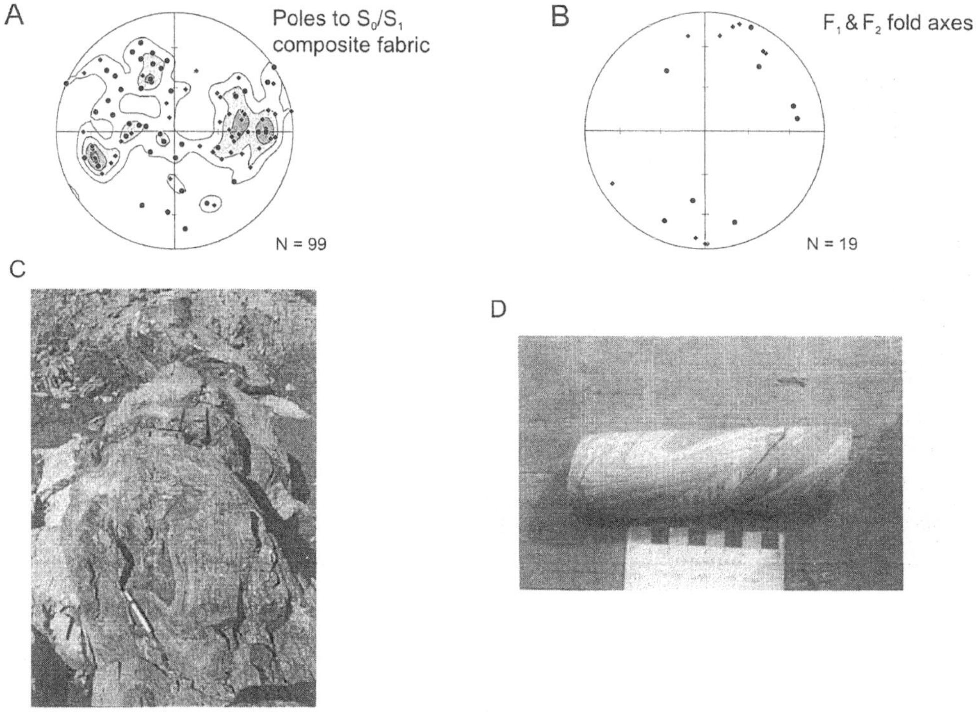

Figure 4. Stereonet plot of poles to the composite S_0/S_1 fabric (A) and measured F_1-F_2 axes (B). For both plots circles represent data from the North Portage area and circles are from the Third Portage area. (C) Photograph of F_2 folds and axial planar cleavage from Third Portage trenchs. (D) Photograph of F_2 folds with a strong penetrative axial planar cleavage from the Third Portage deposit (TP98 291, 38.5).

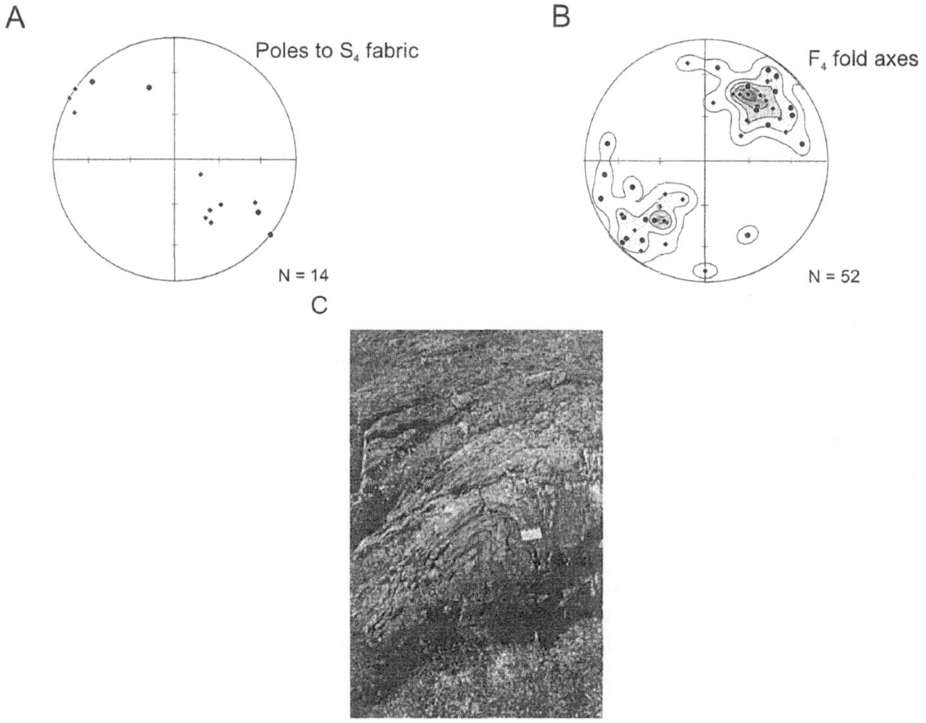

Figure 5. Stereonet plot of poles to S_4 fabric (A) and measured F_4 axes (B). Symbols are the same as Figure 4. (C) Photograph of F_4 fold from the North Portage area.

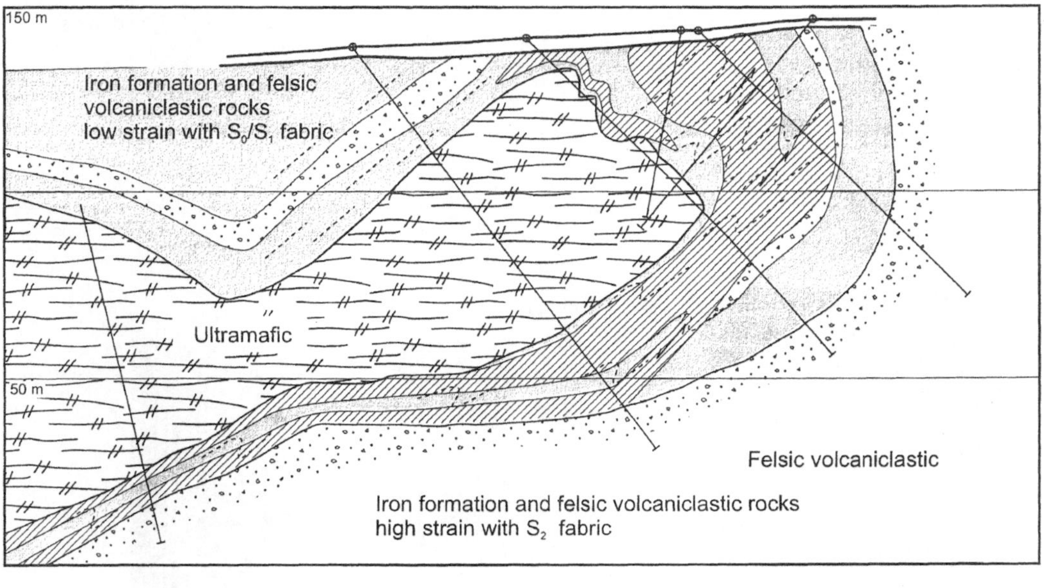

Figure 6. Simplified cross section through the Third Portage deposit at section 1+80N. Dashed lines represent the distribution and asymmetry of deformation fabrics around the ultramafic body. The hatched area represents an interpreted mineralized envelope; any material within the shaded areas will average greater than 1.0 g/t gold.

Gold is contained within sulphide-bearing iron formation (Armitage et al., 1996; Kerswill et al., 1998), characterized by disseminated to semimassive pyrrhotite-pyrite which preferentially replaces magnetite (Armitage et al., 1996). Mineralization is also developed in the felsic-intermediate volcanic package as disseminated pyrrhotite and pyrite along with narrow grey quartz veins also containing pyrrhotite and pyrite.

Sulphides show a variety of relationships with deformation fabrics. Pyrrhotite, pyrite and sulphide-bearing quartz veins can be aligned along S_0/S_1 fabrics as well as within S_2 foliations overprinting S_0/S_1 fabrics (Kerswill et al., 2000; Pehrsson et al. 2000). Spatially, at the Third Portage deposit, the bulk of the mineralization is concentrated at the structural footwall or hinge area of ultramafic units within, and adjacent to, high strain zones defined by penetrative S_2 fabrics. Suggesting either different generations of sulphides or that the sulphides are related to the progressive D_1-D_2 deformation event and concentrated in areas of higher strain.

The degree of transposition of the host stratigraphy makes it difficult to trace individual units within the felsic volcaniclastic and iron formation package. Even in the Third Portage area, where trenching and close spaced diamond drilling provides exceptional data density, the surface trace of the individual units is uncertain. The gold mineralization however is very continuous along strike and down dip. Mineralized envelopes, of greater than 1.0 g/t Au show good continuity and can be traced for 100's of meters. It is this feature of the mineralization which has enabled Cumberland to produce a proven-probable mining reserve. Mineralization occurs in iron formation and volcaniclastic rocks and is more continuous than any individual stratigraphic interval.

Armitage et al. (1996), based on whole rock and mineral chemistry, has shown that auriferous iron formations are characterized by a mineral assemblage of cummingtonite and biotite with pyrrhotite-pyrite-magnetite and quartz. Barren, but sulphidic, iron formations are characterized by an assemblage of grunerite ± hornblende + stilpnomelane with pyrrhotite, pyrite, quartz and magnetite. Whereas barren, non sulfide-bearing, iron formations are characterized by quartz and magnetite. Based on these mineral assemblages and their relationship to gold mineralization, Armitage et al. (1996) have suggested that mineralization is related to metasomatic fluids that have altered the iron formation with the introduction of Mg-K-Ca-S-As-Cu and Au during D_1-D_2 deformation.

Although speculative; it is possible that metasomatic fluids, introduced into dilational sites during D_1-D_2 deformation, altered the iron formation and locally the volcaniclastic rocks. The mechanical contrast between the ultramafic and volcaniclastic rocks partitioned strain into the volcaniclastics and iron formation, which preferentially localized dilational settings. This allowed fluid influx and the resulting alteration assemblages, described by Armitage et al. (1996), to be concentrated in the higher strain corridor adjacent to the ultramafic rocks. The progressive nature of D_1-D_2 formed the various relationships of sulphide minerals and deformation fabrics.

5 ACKNOWLEDGMENTS

The authors gratefully acknowledge the logistical support at the Meadowbank Camp provided by Cumberland Resources Ltd. We also would like to thank Lori Wilkinson, Eva Zaleski and John Kerswill for helpful discussions on the Meadowbank geology. This paper has benefited by a critical review from David Scott and Jan Peter. Comments by Lori Wilkinson, Sally Pehrsson and John Kerswill clarified the presentation and are appreciated.

REFERENCES

Armitage, A.E., James, R.S., & Goff, S.P. 1996. Gold mineralization in Archean banded iron formation, Third Portage Lake area, Northwest Territories, Canada. *Exploration and Mining Geology*, v. 5, p.1-15.

Davis, W.J., & Zaleski, E. 1998. Geochronological investigations of the Woodburn Lake group, western Churchill province, Northwest Territories: preliminary results; in Radiogenic Age and Isotopic Studies: Report 11. *Geological Survey of Canada, Current Research 1998-F*, p. 89-97.

Fraser, J.A. 1988. Geology of Woodburn Lake Map area, District of Keewatin. *Geological Survey of Canada, Paper 87-11.*

Henderson, J.R., & Henderson, M.N. 1994. Geology of the Whitehills-Tehek Lakes area, District of Keewatin, Northwest Territories (parts of 56D, 56E, 66A and 66H). *Geological Survey of Canada, Open File 2923*, scale 1:100 000.

Henderson, J.R., Henderson M.N., Pryer, L.L., & Cresswell, R.G. 1991. Geology of the Whitehills-Tehek Lakes area, District of Keewatin: and Archean supracrustal belt with iron formation hosted gold mineralization in the central Churchill Province. *Geological Survey of Canada, Current Research, Part C; paper 91-1C*, p. 149-156.

Kerswill, J.A., Goff, S.P., Wilkinson, L., Jenner, G.A., Kjarsgaard, B.A., Bretzlaff, R., & Samaras, C. 1998. An update on the metallogeny of the Woodburn Lake group, western Churchill Province, Northwest Territories. *Geological Survey of Canada Current Research 1998-C* p. 29-41.

Kerswill, J.A., Goff, S.P., Kjarsgaard, B.A., Jenner, G.A. & Wilkinson, L. 2000. Highlights of recent metallogenic investigations in western Churchill Province, Nunavut, Canada: implications for mineral exploration in Archean greenstone belts. *[extended abstract] in GeoCanada 2000- The Millennium Geoscience Summit CD-ROM*, Calgary, Abstract 736.

Kerswill. J.A. 2000. Iron-formation-hosted gold deposits: a view from Nunavut with emphasis on "Lupin-like" deposits; [extended abstract with maps] in *Abstract Volume, Short Course on Geology and Mineral Deposits of Nunavut Territory, Society of Economic Geologists, London Student Chapter*, March 3, 2000, 10 p.

Pehrsson, S., Wilkinson, L., Zaleski, E., Kerswill, J., & Alexander, B. 2000: Structural geometry of the Meadowbank deposit area, Woodburn Lake group – implications for a major gold deposit in the Western Churchill Province. *Extended abstract] in GeoCanada 2000, Calgary Alberta.*

Roddick, J.C., Henderson, J.R., & Chapman, H.J. 1992. U-Pb ages from the Archean Whitehills-Tehek lakes supracrustal belt, Churchill Province, District of Keewatin, Northwest Territories. *Radiogenic Age and Isotopic Studies: report 6, Geological Survey of Canada, paper 1992-2*, p. 31-40.

Sandeman, H.A. Brown, J., Studnicki-Gizbert, C., MacHattie, T., Hyde, D., Johnstone, S., Greiner, E., and Plaza, D. 2001. Bedrock mapping in the Committee Bay Belt, Laughland Lake Area (NTS 56K) Central Mainland, Nunavut. *Current Research paper 2001-C, Geological Survey of Canada (in press).*

Schau, M. 1982. Geology of the Prince Albert group in parts of the Walker Lake and Laughland Lake map areas, District of Keewatin. *Geological Survey of Canada, Bulletin 337*, 62 p.

Sherlock, R.L., Alexander, R.B., March, R., Kellner, J. & Barclay, W.A. 2001. Geologic setting of the Meadowbank iron formation-hosted gold deposits. *Current Research paper 2001-C, Geological Survey of Canada (in press).*

Wright, G.M. 1967. Geology of the southeastern barren grounds, pats of the Districts of MacKenzie and Keewatin (Operations Keewatin, Baker, Thelon). *Geological Survey of Canada, Memoir 350*, 91 p.

Zaleski, E., Corrigan, D., Kjarsgaard, B.A., Kerswill, J.A., Jenner, G.A. & Henderson, J.R. 1997a. Geology, Woodburn Lake group, Meadowbank River to Tehek lake (66H/1, 56E/4), District of Keewatin (Nunavut), Northwest Territories. *Geological Survey of Canada, Open File 3461*, scale 1:50 000.

Zaleski, E., Corrigan, D., Kjarsgaard, B.A., Kerwill, J.A., Jenner, G.A., & Henderson, J.R. 1997b. Preliminary results of mapping and structural interpretation from the Woodburn project, western Churchill Province, Northwest Territories; *Geological Survey of Canada Current Research 1997-C* p. 91-100.

Zaleski, E., Duke, N.L., L'Heureux, R., & Wilkinson, L. 1999a. Geology, Woodburn Lake group, Amarulik Lake to Tehek Lake, Kivalliq Region, Nunavut. *Geological Survey of Canada Open File 3743*, scale 1:50 000.

Zaleski, E., L'Heureux, R., Duke, N., Wilkinson, L., & Davis, W.J. 1999b. Komatiitic and felsic volcanic rocks overlain by quartzite, Woodburn Lake group, Meadowbank River Area, Western Churchill Province, Northwest Territories (Nunavut). *Geological Suervey of Canada Current Research 1999-C*

Zaleski, E., Davis, W.J., & Wilkinson, L. 2000. Basement/cover relationships, unconformities and depositional cycles of the Woodburn Lake group, western Churchill Province Nunavut. *[abstract] Yellowknife Geoscience Forum, November 2000.*

Government administration and legislation related to mining

Mining in the Arctic, Olsen, Lorentzen & Rendal (eds), © 2001 Swets & Zeitlinger, Lisse, ISBN 90 5809 177 5

Dealing with failure of Royal Oak Mines in the Northwest Territories

R. W. Lauer
Department of Indian Affairs and Northern Development, Hull, PQ, Canada

D. E. Nutter & N. A. Thompson
Department of Indian Affairs and Northern Development, Yellowknife, NT, Canada

ABSTRACT: Royal Oak Mines Inc. was placed in receivership on April 16, 1999, and, on December 14, 1999, the Canadian government was left to deal with two abandoned mines in the Northwest Territories. Colomac, a 10,000 ton per day, low grade, open pit gold mine on care and maintenance since 1997, had a tailings management problem resulting from the accumulation of cyanide-bearing effluent, as well as geotechnical problems with the retention dams. Giant, a 1000 ton per day, underground gold mine operating since 1948, processed its arsenopyrite ore through an outdated roasting facility, resulting in airborne arsenic trioxide and sulphur dioxide emissions and generating 265,000 tons of arsenic trioxide-bearing dust currently stored in underground chambers. The Department of Indian Affairs and Northern Development must manage potentially significant environmental, human health and socio-economic impacts under the legal and financial constraints imposed by insolvency legislation and within an atmosphere of close public scrutiny.

1 INTRODUCTION

Royal Oak Mines Inc. was a mid-sized mining company with precious and base metals properties across Canada, including two gold producing operations in the Northwest Territories: the Colomac and Giant mines.

By early 1999, Royal Oak had amassed debts in excess of Cdn $600,000,000, primarily as a result of the development of a large base and precious metals property in British Columbia during a time of weakening metal prices. On February 15, 1999, Royal Oak was granted court protection from its creditors, and, on April 16, 1999, the Ontario Court of Justice appointed PricewaterhouseCoopers Inc. as Interim Receiver for all of Royal Oak's assets. The Interim Receiver assumed responsibility for carrying on the business and affairs of Royal Oak, with the objective of reorganizing the company or liquidating the assets for the benefit of the creditors.

In the Northwest Territories, as in Canada's other northern territories, most public land is administered by the federal government through the Department of Indian Affairs and Northern Development (DIAND). DIAND is responsible for the administration of lands and water and the disposition of mineral rights; in this sense, playing a role similar to that of a provincial government in southern Canada.

2 BACKGROUND

2.1 *Colomac*

The Colomac mine, a 10,000 ton per day (tpd), open pit gold mine located 200 km north of Yellowknife, was placed on care and maintenance in December, 1997. While the mine only operated for approximately three and a half years during its decade of existence, there are significant environmental issues to be addressed.

By early 1999, the property was at risk of releasing toxic concentrations of cyanide into the local drainage system. The tailings containment area was designed as a zero discharge facility; however, it became apparent that the tailings pond had a positive water balance. A controlled release of tailings fluid will be necessary, but cyanide, copper and ammonia levels must first be reduced. In late winter, 1998-1999, it was determined that, unless immediate actions were taken, there would be an uncontrolled release of tailings effluent during the spring runoff. Royal Oak staff were able to avert a serious problem by transferring excess effluent to an abandoned open pit, but the threat of an uncontrolled release has been a recurring problem, and storage in the open pit is only a temporary solution.

Soil testing in the vicinity of the site's fuel storage tank farm indicates that there has been substantial leakage during the life of the mine. There is severe hydrocarbon contamination within both the overburden and the underlying bedrock. In addition, there are numerous hazardous and non-hazardous waste sites about the property, which must be reclaimed.

2.2 *Giant*

The Giant mine, a 1000 tpd, underground gold producer, located in Yellowknife, has been operating for more than 50 years. Until October, 1999, the operation relied on roaster technology in its gold recovery process, and was producing up to 8 tons of byproduct arsenic trioxide per day. Since the early 1950s, more that 265,000 tons of arsenic trioxide-bearing dust has been recovered from the roaster and stored in underground chambers at the mine.

Sampling of mine water during the early 1990s revealed that arsenic trioxide was being leached out of the storage vaults. Sumps and the mine's dewatering system were capturing this contaminated water, and it was pumped to a water treatment plant on surface. However, the acceptability of this as a long term management solution began to be questioned.

Moreover, after more than a half century of operation, the Giant property was a heavily contaminated industrial site. As a fifty year old mine where progressive site reclamation had not been practised, the Giant mine posed many environmental challenges. Arsenic and hydrocarbon-contaminated soils, asbestos, abandoned materials and equipment, several large tailings containment areas, and fuel and waste oil storage sites, plus many industrial and residential structures, were found throughout the property, and all would need to be dealt with.

3 CHALLENGES AND SOLUTIONS

In this context, the insolvency of Royal Oak posed a number of significant environmental, financial and regulatory challenges for the federal government. The environmental liabilities at the Colomac and Giant properties far exceeded the value of the assets at both mine sites, including reclamation security deposits which had been posted by Royal Oak with respect to the two operations. As a result, it was not surprising that the Interim Receiver was unable to find a buyer interested in acquiring either property on an as-is-where-is basis. In the absence of a purchaser, the Interim Receiver might eventually have departed from these properties without an orderly transfer to another party.

From an operational perspective, DIAND was relatively well-prepared to deal with these properties. From early 1999, when it appeared that the Colomac and Giant properties could be abandoned at any time, the department had contingency plans in place and had retained qualified contractors who were prepared to step in and manage these operations on short notice. At the end of 1999, the real challenge for DIAND was minimizing the costs to the Canadian government of safeguarding public health and safety and protecting the environment at these properties.

In this situation, DIAND had the advantage of relatively recent amendments to federal insolvency legislation which gave the Government of Canada a first charge over the real property assets of mines such as Colomac and Giant to cover the costs of environmental remediation. If such a property was abandoned and if Canada stepped in to protect the environment, its expenditures would become secured by this charge. This allowed DIAND to determine how these properties were dealt with in the court process and enabled DIAND to extract any residual value from these assets ahead of all other creditors. Given the federal government's first charge over each property, in December, 1999, the Interim Receiver transferred all of the assets at both properties to DIAND.

How DIAND dealt with each property was a reflection of their individual situations. A significant difference between the two properties was that, under existing market conditions, Colomac had no mineable reserves, but there was still some mineralization at Giant that could be considered to be economically recoverable. As a result, DIAND's approach to dealing with each property was radically different.

3.1 *Colomac*

In retrospect, dealing with Colomac was definitely the more straight-forward of the two. Prior to the court-approved transfer, DIAND had made arrangements with the Interim Receiver for an orderly transition of operational responsibility for the Colomac site. This is a remote property, accessible only via aircraft and winter road, and there was strong desire for an orderly transfer before the onset of winter. The Interim Receiver wished to divest itself of the property and was reluctant to incur remediation expenses. On the other hand, DIAND believed that site reclamation should begin as quickly as possible. Furthermore, assuming operatorship would enable it to claim reimbursement against the mine's security deposit. DIAND had been working closely with Royal Oak staff, and the department was fairly familiar with the operational demands of the property.

DIAND continues to manage the property, utilizing

northern contractors. When Royal Oak suspended operations at Colomac in 1997, most salvageable equipment was removed from the site. DIAND has had to operate with machinery which requires constant maintenance and attention, and the challenges of day-to-day operations have been substantial

The property is being kept in environmental compliance, and reclamation will proceed as funds become available. The department will eventually sell whatever assets it can to help defray some of the cost to the Canadian government of environmental remediation.

Under the Colomac Water Licence, Royal Oak had posted a Cdn $1,500,000 security deposit. While DIAND, in its regulatory role under the Northwest Territories Waters Act, has a claim on these funds, it was believed to be prudent to first incur reasonable expenditures in order to realize this claim. The department has advised the insurer of its claim on the performance bond. However, the total cost of mine reclamation, currently estimated at Cdn $16,000,000, will far exceed the value of the security deposit and the proceeds of the sale of assets that were transferred to DIAND.

3.2 *Giant*

The situation at Giant was considerably more complex. In contrast to the Colomac mine which was inactive throughout 1999, Giant was in full production at a rate of approximately 1000 tons per day, with over 300 workers, until operations were shut down by the Interim Receiver in October, 1999. However, there remained mineral reserves at Giant which could conceivably be economically mined, and, therefore, there was intense pressure from the local community of Yellowknife for a solution which would preserve employment. At the same time, there was significant local concern about the environmental impact of the mine's operations. While Giant's roaster technology was fully licenced, it was producing up to 8 tons of arsenic trioxide-bearing dust daily, which was placed in the underground storage vaults. In addition, the levels of airborne arsenic trioxide and sulphur dioxide emissions were cause for concern within the community.

The situation at Giant has been further complicated by a number of other factors.

- The federal government has transferred the administration of the surface lands within the City of Yellowknife, including the Giant mine site, to the Government of the Northwest Territories (GNWT).
- The GNWT maintains that responsibility for remediation of the Giant site was not included within the transfer. The result has been a disagreement between the federal and territorial governments as to which one is liable for the costs of surface reclamation at Giant.
- The City of Yellowknife maintained that any buyer of the Giant mine must assume Royal Oak's unpaid municipal tax liability.
- Royal Oak's insolvency left the pension plan for Giant's hourly workers in a deficit and severance packages unpaid.

In the middle of this complex set of conflicting legal, financial, socio-economic and environmental pressures, DIAND decided on a solution which had no precedent that the department was aware of.

When the Interim Receiver was looking for a purchaser of Giant, several offers were received to take over operation of the Giant mine as an agent for the Government of Canada. This was not an acceptable solution to the federal government, which was averse to owning and operating a gold mine.

However, one mining company, Miramar Mining Corporation (Miramar), which owns the adjacent Con mine in Yellowknife, indicated a willingness to take over the Giant Mine, as long as it was insulated from the existing liabilities associated with the mine. Miramar proposed to cease operation of the Giant mill and roaster and to ship the ore from the Giant operation to Miramar's neighbouring Con mine for processing through its mill and autoclave. This would eliminate a major source of pollution in Yellowknife and, at the same time, preserve some employment for the laid-off mine workers.

On this basis, DIAND and Miramar negotiated a series of agreements under which the mine would be transferred to Miramar, and, when the Interim Receiver transferred Giant to the Crown in December 14, 1999, the mine was sold to Miramar Giant Mine Ltd. (Giantco), a subsidiary of Miramar. There were several key provisions to these agreements.

- Giantco would maintain the property in environmental compliance.
- Giantco would have the right to mine throughout the Giant property.
- Miramar's liability for the environmental obligations related to the state of the mine at the time of purchase would be limited to the assets of Giantco.
- Profits from Giantco's operations, including proceeds from the sale of its assets, would be shared with a reclamation security trust, established to fund Giant reclamation.
- Giantco committed to hold the Giant property for a minimum of two years and to provide DIAND with six months notice of its intention to abandon the property.

DIAND was attracted to the deal with Miramar for a number of reasons.

- With the imminent abandonment of the mine by Royal Oak, DIAND considered it to be in the public interest to retain a qualified and experienced operator for this complex facility.
- It was estimated that the cost of maintaining the Giant facility in environmental compliance, pending the identification of a long term management solution for the arsenic trioxide, would be Cdn $2,000,000 to $3,000,000 annually. Giantco assumed this liability for at least two years and relieved the federal government of the immediate financial and managerial pressure of maintaining the Giant property in environmental compliance.
- While the Giant mine is now operating at a reduced rate of approximately 350 tons per day, the continuation of operations has resulted in the ongoing employment of approximately 60 former Royal Oak employees.
- No further arsenic trioxide or sulphur dioxide will be produced by the Giant mine.
- The sale and continued operation of Giant will also provide some funds for site remediation.
- This sale has provided DIAND with additional time to identify the most promising solution for the long term management of the 265,000 tons of arsenic trioxide stored underground at Giant.

Miramar gained an additional source of ore for its Con operation. The company also enhanced its involvement in the development and implementation of arsenic trioxide remediation efforts in the community, which is of interest to Miramar because the Con mine also hosts significant quantities of arsenic trioxide stored in licenced, surface compounds.

While, from a strictly legal perspective, Miramar purchased the Giant mine from the Canadian government subject to certain conditions, the reality is that the agreements have resulted in a unique public/private sector partnership where both parties need and rely upon the other. Through its Water Licence, Giantco, as the operator of the mine, is legally responsible for the environmental condition of the Giant property; however, DIAND has assumed the burden of coping with essentially all of Giant's pre-existing environmental conditions. Miramar is preparing a reclamation plan for the surface of the mine and has provided DIAND with invaluable advice and assistance in analysing the various long-term remediation options for the arsenic trioxide stored underground in the Giant mine.

Adding to the complexity of managing the complete remediation of the Giant property, the City of Yellowknife is in the process of acquiring the lands formerly occupied by the Giant town site. The City intends to develop the recreational potential of this waterfront property, and agreement must be reached between the federal, territorial and municipal governments regarding their individual reclamation responsibilities.

Under the Giant Water Licence, Royal Oak had posted a Cdn $400,000 security deposit, far below the actual cost of full reclamation. In conjunction with its Colomac claim, DIAND has filed a claim against the insurer for partial reimbursement of its Giant costs.

While DIAND's deal with Miramar and its efforts to realize on the security deposit do reduce the present day cost to the Canadian government of environmental protection at the Giant Mine, they would not begin to offset the cost of complete environmental remediation, which has been estimated at up to Cdn $390,000,000 depending upon the solution adopted.

4 OBSERVATIONS AND LESSONS LEARNED

Over the past decade, DIAND has been forced to deal with a number of mining industry insolvencies involving old mines where environmental liabilities exceed the security held by the Crown: Curragh, Anvil Range Mining, BYG Resources, United Keno Hill and Royal Oak Mines. While each situation is unique, some common principles have become apparent, and, by making use of this experience, DIAND is better prepared to deal with each new situation.

When a mine operator is becoming insolvent, it is critical for the environmental regulator to be fully familiar with the physical situation at each affected site and to have contingency plans in place for all anticipated scenarios. The regulator should have a good knowledge of both insolvency and environmental legislation and hire top notch legal counsel with expertise in commercial and insolvency law to represent the regulator in court, where many of the critical decisions are made. It is also important that the regulators in the field work closely with the legal and financial expertise within their organization.

Through these experiences, DIAND has learned the importance of taking the initiative in the legal process, sometimes aggressively, to influence the course of events and decisions in such a way that the interests of the public - both financial and environmental - are protected whenever possible. Insolvency is a zero-sum game. By definition, there is not enough to satisfy all of the claims against an insolvent mine operator. Any party, including the Canadian government, which does not actively defend its interests in this process will come up very short.

At the same time, the importance of being flexible when dealing with environmental liabilities is readily apparent. In the case of Giant, this was what opened the door for a private sector mining company, Miramar, to assist Canada in finding a solution to a very challenging problem.

In a number of instances, DIAND has begun as solely the regulator but, eventually, found itself in the position of de facto owner. This creates inherent challenges for any organization, where it ends up regulating itself. Different parts of the organization must, at the same time, be able to work collaboratively towards the same goals - protection of public health and the environment - while maintaining a sufficiently arms length relationship to avoid conflict of interest situations.

In many ways, the situation at Giant highlights one which commonly exists in mining jurisdictions, namely: old, high-cost mines that were developed prior to the adoption of current environmental standards. Frequently, the costs of environmental remediation at these mines exceed their reclamation security, and any attempt to substantially increase financial security will simply drive the operator out of business, leaving government with the problem. There is no easy solution to this situation. To prevent this type of problem in the future, government may need to consider much more substantial financial security than in the past. Recent examples of this policy in the Northwest Territories include the cases of the Ekati Diamond Mine operated by BHP Diamonds Inc. and the new Diavik Diamonds Project being developed by Diavik Diamond Mines Inc., a subsidiary of Rio Tinto plc, and Aber Resources Ltd.

5 CONCLUSIONS

1. Know and use your legislation. Understand the art of the possible within the existing regulatory framework.

2. Be pro-active. Aggressively defend the public's interests and lead the way in finding solutions.

3. Be creative. Flexibility and an open mind often bring solutions which are the most efficient and cost effective to implement.

4. Use the law. Strong legal counsel is a prerequisite. In insolvency law, the most aggressive party has the upper hand.

5. Anticipate. Develop response and communications plans for all potential scenarios, and establish a project team.

Mining in the Arctic, Olsen, Lorentzen & Rendal (eds), © 2001 Swets & Zeitlinger, Lisse, ISBN 90 5809 177 5

Non-geological issues pertaining to prospect evaluations

A. Orheim
GeoArktis, Longyearbyen, Norway

S. E. Sørstrøm
GND Arcticon AS, Longyearbyen, Norway

ABSTRACT: The Arctic regions represent industrial frontier areas. Large parts are public land and other parts belong to the indigenous people. To enhance development mining regulations are lenient, e.g. tax incentives or large mining acreage. Environmental awareness may change this, and require impact assessments prior to future development. Svalbard represents a unique situation: No indigenous inhabitants; Neither private nor public land prior to 1925; Until recently mining was the sole all-year commercial activity; Prospect evaluation has been decided by economy and environmental issues, although international politics should also be credited. Presently Store Norske is developing a new mine at Svea. Comparatively, this may be one of the most economic and environmentally acceptable coal mines on a global scale. Despite the economic an environmental benefits, there are still numerous discussions whether coal mining is acceptable in highly specialized and sensitive ecosystems like the Arctic.

1 INTRODUCTION

We assume it is acceptable to observe the Arctic as one generic region regarding commercial interests. A global study by the United Nations (Otto et al. 1992) lists 9 different business criteria applied by the industry when considering mining propositions. This paper highlights some of the criteria most applicable to the Arctic region, and we shall discuss how they have influenced resource assessments and development of Store Norske's mine at Svea, Svalbard.

The UN study summarized the following 9 criteria: Geology, Policy, Market, Regulatory, Fiscal, Monetary, Environment, Operations and Profit. They are all of importance wherever and whenever an investment is contemplated. However, it is our opinion that issues related to the environment, policy and regulatory conditions have a decisive impact when a company is considering mineral deposits in the Arctic.

2 GENERIC CONDITIONS OF THE ARCTIC

2.1 *Regulatory matters*

Ownership to mineral resources is one key issue in mineral legislation. 2 different judicial views have endured through 2000 years. In ancient Greece mineral resources were considered belonging to the State regardless of ownership to the land surface, while the Romans acknowledged ownership of extractable minerals to the owner of the land under which the resources could be found.

The Latin view came to be the guiding principle of Western Europe, notably France and England, while a more Hellenistic view may be traced to the North e.g. Germany and Scandinavia. Here all land except areas just underneath houses, were considered public land, open to common use and where anyone could make use of the resources. As the structure of nations developed, landlords and royalty got the upper hand and took charge of the public land; It became part of the Regalia, under the King's disposition. And in most cases the king would leave the land free to prospectors. (Helland 1892).

The colonial powers introduced their legal system in the "new" lands. Hence in Greenland the Danish king had monopoly to all mineral extraction. In New England, the owner of farmland also owned extractable subsurface resources.

As migration moved westwards in North America, fuelled by dreams of gold and silver, the prospectors moved onto unclaimed land, i.e. still public land free to all. In these regions the mining regulations became more similar to the ancient Hellenistic view: every citizen is free to make mineral discoveries and claim the rights to extraction. And, the person holding the oldest discovery is given priority.

Regional variation of the US confirms the view that a mining code based on the ancient Greek principles is best adopted in regions of low level of urbanization, e.g. the Arctic. This code will be in favour of the innovative prospector.

2.2 *Political issues*

There is a close relationship between community development, industrial infrastructure, public installations and various political issues. And of course, this relationship is particularly obvious in the Arctic. Political questions are both of international scale (cold war effects) and of a more domestic nature related to the particular regions.

The Arctic is one of the last urbanized regions of the globe and community development would normally not be achieved without a political engagement outside of the region itself. The willingness to source various scheme of development has improved as awareness of the questions related to indigenous people have risen. Thus the mining community might be faced with questions regarding justified extraction of resources or they could fine economic incentives to develop geological resources and local industry.

This is exactly what happened regarding exploration of hydrocarbon in the Norwegian Barents Sea. In 1980 the first well was drilled in the Barents and next year saw the first discovery. By 1993 53 wells had been completed, including 16 discoveries. All of them too small to justify further development, and gradually the industry lost interest. No wells were completed in the following years.

The politicians in Norway realized a new regulatory system had to be established. Larger acreage and provision for the companies to make joint (group) application were the key measures. They allowed for a renewed activity, and of 4 wells completed last year 3 gave discoveries, including the first major oil discovery, Figure 1.

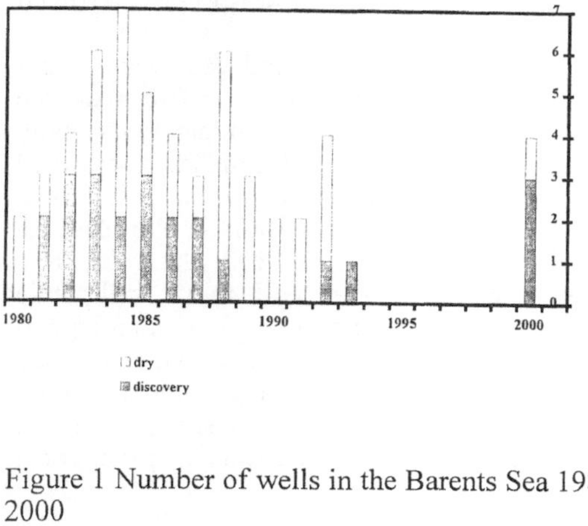

Figure 1 Number of wells in the Barents Sea 1980 – 2000

2.3 *Environmental issues*

Through the latter part of last century the environment gradually became the common global concern. Polar ecosystems are vulnerable to environmental changes. Thus, some may regard mining in the Arctic as contradictory to any acceptable development. However through the planning and development of the Svea mine it has been demonstrated that environmental issues may act as guiding principles of every activity, and serve to focus management attention on achievable environmental goals.

In a modern society it is impossible to imagine a situation without refining, storage, distribution or use of energy. The economy of the world society enforces a constantly increasing supply of energy, with the consequences that energy supply becomes the most important global environmental challenge. According to the World Energy Council, the distribution between different sources has been relatively stable in recent years. As a consequence, coal will still be the most important source of energy for decades to come.

2.3.1 *Coal and the environment*

Coal contains a number of mineral components in addition to carbon and hydrogen. Combustion, gasification or pyrolysis may release these components into the flue gas. SO_2 and NO_x, which are released or produced during combustion, create acids in the atmosphere that may return to the ground as acid rain. Air pollution affects the environment and health of millions of people, in particular in the larger cities.

2.3.2 *International collaboration*

Norway has a very low level of coal consumption – most of her energy usage comes from hydropower. At the same time Norway is the only producer of coal in the Nordic countries and is also one of the largest exporters of oil and gas. Norwegian economy is to a large extent based on export of fossil fuels.

On the other hand, China has the highest dependency of coal in the world. The growth in coal consumption combined with high economic growth and low access to modern technology will cause tremendous increase in emission of local air pollution and green house gasses. As a result of the extensive use of coal for energy purposes, the areas covering south-west and south China have become one of the three major global regions regarding acid rain, and China ranks third in terms of emissions of carbon dioxide.

One possible benefit of the adopted system for impact assessment employed at Svea, may be to exchange experiences with the Chinese, and support their development of similar, and equally applicable systems for environmental impact assessment.

3 MINERAL INVESTMENT IN SVALBARD

3.1 *The mining code*

Through 3 centuries of whaling, hunting and scientific field work (1600-1900) Svalbard remained a land with no formal jurisdiction, a *terra nullius*. However, when coal and other mineral resources attracted prospectors, a legal framework became a necessity. In 1906 the American J. M. Longyear started coal mining in Longyearbyen. Store Norske bought the property in 1916, and 3 years later, the Norwegian representative to the Versailles conference following World War 1, was able to bring forward Svalbard as an issue pertaining to international law.

In 1920 a treaty could be signed giving full Norwegian sovereignty, although with certain provisions, namely: Free access for every citizen of the 41 signatory powers to start virtually any kind commercial enterprise; Established interests should be granted title to necessary land; Restricted level of taxation; No fixed military installations; Provisions to ensure endangered species; and finally: Norway should propose a mining code to the signatory powers prior to enforcing the treaty. The treaty and mining code was finally sanctioned by 1925.

The mining code provides legislation regarding how to ensure and maintain mining rights and specifies requirements regarding safe operations and the health care of employees. To ensure mineral rights the interested party will have to physically mark the outcrop of a commercial viable resource and collect a sample at that particular location. Within 10 months the find must be documented to the Commissioner of Mines. Provided he finds everything in order, and no one else has an older discovery in the same area, there is a certainty that a claim of 10 km^2 will be granted. The claim holder may extract every kind of commercial mineral within that area. The mining code stipulates certain fixed obligations: time limits, minimum level of activity and dues. However, as long as you abide to all of these you will have your undisputed mining rights.

By and large this is similar to other mining codes. However, as Svalbard was no part of any country when mining started, prior to 1925, the mining code could not provide for any principle of land ownership (or Regalia) to Norway. Neither the basic assumption of the Roman nor the Hellenistic view regarding mineral resources is applicable at Svalbard. The land owner is entitled to a 25% participation, however he is not entitled to have an influence on the actual management of the mine. Thus, compared to most other places where the miner would have to apply for license from the authorities and/or the land owner, the claim holder is in a more independent position. Compared to other modern, complicated and organized societies this leaves Svalbard in a relative secluded position.

On the other hand, as the treaty also instructs Norway to maintain the fragile ecosystems, large parts of the archipelago are now national parks, i.e. out of bounds to mineral exploration. And any surface installation has to meet the criteria provided by the Norwegian Planning and Building Act or similar institutions. As these regulations are enforced equally to everyone the basic principles of the treaty remain undisputed.

3.2 *Immediate Experiences*

The mining code of Svalbard provides for simple and just treatment of every mineral prospector. This has attracted both serious and more speculative investments. Looking at mining activities may give some valid experiences regarding how mineral investments develop.

The Svalbard Treaty also gave rules regarding how to assign ownership of land to interests already in place prior to 1925. Entitled parties should document their demand towards a Danish, i.e. neutral, commissioner. He was empowered to conclude in all matters, through arbitration if necessary, and Norway should accept his ruling.

The commissioner received 74 demands from 26 different claimants. Gold, petroleum, asbestos, marble, calcite, coal, galena, chalcopyrite, sphalerite and numerous others minerals were listed as having possible commercial interest. Several of the demands were in conflict and he decided by assigning 49 properties to 15 commercial interests. 10 of the interests were Norwegian but the British got the majority of properties, 20.

The timing of the treaty was not favourable. Within a few months after 1925 the post-war economy started to collapse, and consolidation of properties became the order of the day. After a few years the number of commercial interest was reduced to 9, of which the Norwegian State and a Russian state operated coal company, were totally new players.

We may make the following observations regarding the enforcement of the Treaty in 1925:

The treaty was meant to provide for law abiding conditions for mining activities, and a dynamic system where geological and commercial changes would be accepted.

The number of property claims to the commissioner confirmed that this was achieved.

The acceptance of the commissioner's ruling also showed that everybody considered that ruling just.

The change from 15 to 9 property owners (including 2 new ones) showed that the treaty and mining code would provide for a dynamic regime.

Both the Treaty and the Mining Code have been left virtually unchanged. This has given the industry a stable working environment and enhanced mining propositions.

3.3 *Experiences through the last 4 decades*

One of the key concerns was to allow for a simplistic judicial process. Business decisions should be left directly with the commercial interests, whether they being serious operators within mining or more speculative operators at the fringe of geological exploration.

In 1960 the American joint venture Caltex made the first move towards mapping of petroleum resources. Within a few years the Commissioner of Mines reported a quadrupled increase of valid mineral claims. However, as the first well, and later wells both by Caltex and others, turned in dry, both the serious and the speculative players lost interest.

Looking at the distribution of mineral claims (Figure 2) the development sparked by Caltex may easily be detected. According to the mining code mineral claims will normally be assigned 5-7 years after a discovery is staked in the field, and 4 years after assigning the claim an obligation to do a minimum amount of work will come into effect. However the work is only reported at the end of consecutive 5 year periods. In other words, the first report will be 9 years after a claim became valid. Looking at the figure we see that some years after the first well was drilled, 1966, there is sharp rise in number of valid claims. And then, 9 years after the well turned in dry the other claim holders starts to relinquish their commercial interests.

A similar development is true in the 1980s, this time fuelled by British Petroleum. In 1984 and 1985 they made large seismic acquisitions on land. Within a few months, literally, both geologic mapping, seismic acquisitions onshore and offshore and renewed drilling programs were started. BP, having come to Svalbard with a clear commercial objective, soon lost interest when they realised that their play model did not match realities. However many of the other companies, without the same specific geological understanding (?) stayed on for several years.

3.4 *Exploration for the new Svea Mine*

Figure 3 is a map of Longyearbyen and Svea, the 2 mining locations of Store Norske, 50 km apart. In the eastern part of the map the Tertiary coal outcrops 50-400 meter above sea level, while further west it is below sea level. The main area of mining from 1906 till 1999 was Longyearbyen, with smaller additions coming from mines in Svea. In addition the company has had a specific exploration programme annually for the last 30 years.

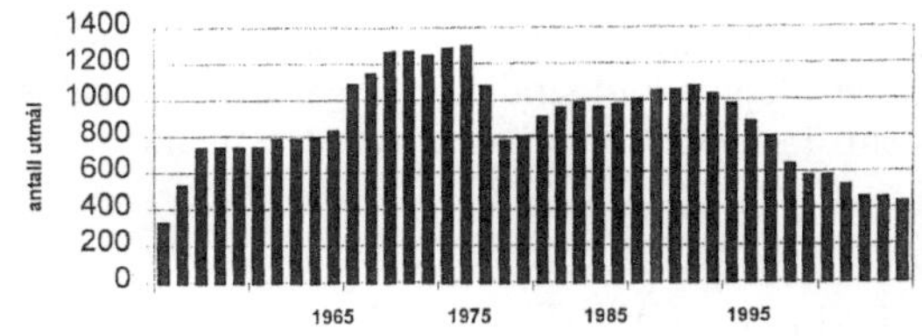

Figure 2 Number of mineral claims in Svalbard 1961 – 2000

A resource assessment by Store Norske in early 1980s (Orheim 1982) gave a 50% probability to identity more than 20 million tons extractable coal given a continous mapping at the same level for the next 30 years. At that time a 5 meter coal seam was being worked in parts of the old Svea mine, while a seam of less than 2 meter was worked at Mine 7, Longyearbyen. Surface mapping had proven a seam of 3.2 meter thickness at Lunckefjell, 17 km north of Svea.

Figure 3 Store Norske mining area Longyearbyen - Svea

Through drilling further west in Reindalen the seam was identified at 1.7 meter in 1984 and the company embarked on a larger drilling program across the glaciers towards Svea. Due to environmental restriction the drilling could only be performed when the ground was frozen, and necessary skills to tackle winter condition had to be developed. The attitude: "don't fight the winter, play along with it".

The first major drilling result was 4 years later when 5 meter coal was encountered in the northern parts of the new mining area. Later more than 70 drill holes penetrated the seam giving us a fairly comprehensive understanding of the geology and necessary information for planning the mine layout.

More than 30 million tons extractable coal have been identified through less than 20 years of exploration, far exceeding the prognosis of 1982.

4 IMPACT ASSESSMENT ANALYSIS

4.1 *Coal dust*

The Environmental Impact Assessment prior to opening of the new Svea coal mine has revealed a number of areas where environmental improvements are possible. One example is discharge of coal dust. The loss of coal dust is approximately 25,000 tonnes annually when the total production is 1.2 million tonnes of coal pr year. The economic loss caused by coal dust amounts to NOK 5 million (US$ 600,000).

Transportation of coal is carried out by large trucks from the mine to the harbour. The cost of road transportation has initiated discussions whether the company should invest in an underground conveyor belt from the mine to the harbour, a distance of 15 km. The total cost for transportation by truck over a period of 20 years amounts to approximately NOK 100 million while investment in conveyor belt will be NOK 122 million. However, the costs both for maintenance and operations is assumed to be much higher for the road compared to a conveyor.

By introducing conveyor belt we may, as an example, assume that the discharge of coal dust will be reduced by 50%. This reduces the economic loss by NOK 2.5 million per year or NOK 50 mill over a period of 20 years. Thus the investment in conveyor belt may be compatible in economic terms by the investment and cost by operating trucks over the same period.

4.2 *Methane and carbon dioxide*

The emission of methane from Svea Nord is measured through a number of different methods. The average emission is 0.8 m^3 methane pr tonnes coal compared to values between 4 and 60 m^3/ton from other coal mines supplying coal to the European market.

Emission of carbon dioxide from transportation of coal to the market is calculated. The total emission related to production and transportation of 1 million tonnes coal to the European market is 43,525 tonnes CO_2 as compared to figures between 144,788 and 709,340 tonnes from other suppliers to the European market. If CO_2 tax is implemented on a global or European scale this will strengthen the value of Svea coal compared to coal from other suppliers to the European market. Coal from markets far from the European consumers will envision competitive disadvantage compared to low emission coal from Svalbard.

The environmental impact analysis has formed an important part of the company's strategy; the ambition to becoming one of the most environmentally friendly coal mines in the world. Taking the correct actions will not enforce additional costs on the company, but rather improve the economic compatibility and strengthen the competitiveness of the Svea coal in the European market.

On the other hand; the environmental cost-benefit ratio is much more favourable in countries like China than in industrialized countries. Outdated technology combined with lack of proper knowledge increases the environmental problems related to production and use of coal for energy and industrial purposes. Knowledge developed through coal mining in sensitive areas like the Arctic may be transferred to areas where the needs for environmental friendly solutions are more obvious.

5 SUMMARY

The Arctic is one of the least urbanised areas of the globe. To achieve sound development of mining in these areas, we believe a mineral legislation open to any prospector on equal terms, has proven to give the optimum result. In order to attract the mineral investor it is important to have fixed mining regulations regarding extraction, safe working conditions and environmentally sound operations.

The Svalbard Treaty and the Mining Code was institutionalised through international negotiations in 1920. The object was to give a judicial framework for mining and other commercial activities. At the same time Norway was granted full jurisdiction. The Mining Code is a consequence of an international treaty. Thus, it has been left virtually unchanged for the last 75 years. This has proven favourable to the mineral investor.

One key provision in 1920 was to ensure equal rights to everybody, and at the same time allow for a system enforcing certain obligations to anyone seeking mining prospects or holding mining claims. Looking at the mining history we have shown that the judicial system has worked exactly as planned.

The exploration and development activities of Store Norske regarding the new mine at Svea has also demonstrated that it is possible to complete these activities within the framework of present environmental concerns.

REFERENCES

Helland, A. 1982. *Norsk Bergret.* Kristiania: Aschehoug.

Orheim, A. 1982. Undiscovered Tertiary Coal Resources of Svalbard: An assessment using Monte

Carlo Simultation. *Arctic Geology and Geophysics. Canadian Society of Petroleum Geologists. Memoir8:* 399-413

Otto, J. 1992. Criteria for Assessing Mineral Investment Conditions. *Mineral Investment Conditions in Selected Countries of the Asia-Pacific Region:* 6-34. United Nations, New York.

Mining in the Arctic, Olsen, Lorentzen & Rendal (eds), © 2001 Swets & Zeitlinger, Lisse, ISBN 90 5809 177 5

Decommissioning shallow mine workings in permafrost terrains

M. C. Bétournay & J. E. Udd
Mining Laboratories, Natural Resources Canada, Ottawa, Canada

ABSTRACT: Shallow stopes of inactive hard rock mines have regularly been subject to failures. Over time, site information is lost and, until recently, no closure plans have been routinely fomulated. A national Standard has now been issued to support the decommissioning of shallow stopes. Taking into consideration the aspects of mines in permafrost terrains, it is presented as an overview here, in the areas prescibed to arrive at long-term stability assessment and related risk and solutions: information management, engineering stability, site management and engineering solutions. To date, there has been a limited number of abandoned mine workings in the Canadian arctic. With increasing minerals development activity, however, the problems need to be addressed. Abandoned mines represent non-traditional hazards to local people.

1. INTRODUCTION

Over the last several years, the regular occurrence of inactive mine shallow stope failures has gained the attention of the mining industry and civic officials. Shallow stopes are defined as the underground mine openings closest to surface, usually within 30 m [Bétournay, 1986], Figure 1. They are normally affected by difficult rock mass conditions (high joint density, wide joint apertures, rock alteration, lower confining ground stresses, etc.) Which can benefit from the positive action of ice filling spaces in the rock mass.

The infrastructure and mine workings that remain today are both a reminder and a legacy of this activity. Many are hazardous and most lack long-term closure plans to address their management. Unfortunately, until recently, few mine plans and drawings have survived mine closures. They have been lost or have deteriorated over the years. Often, a mine site no longer has surface structures to mark its location and the presence of underground mine openings is unknown. In the north, abandoned mine workings can soon be covered by snow and may represent hazards to local inhabitants.

Long after mines have closed, the local population and governments consider mine hazards to be limited to those in full view. Consequently, people undertake activities without being aware of underground rock mass site conditions and historical failures. Furthermore, in the absence of a national building code which requires assessments of geotechnical risk potential, many mining towns, infrastructure, or transportation arteries (highway, rail) have been built above inactive shallow underground openings.

Closure of mines in Canada has only recently required a-priori planning to evaluate and mitigate long-term rock mass degradation, permafrost thawing and potential for rockmass movements

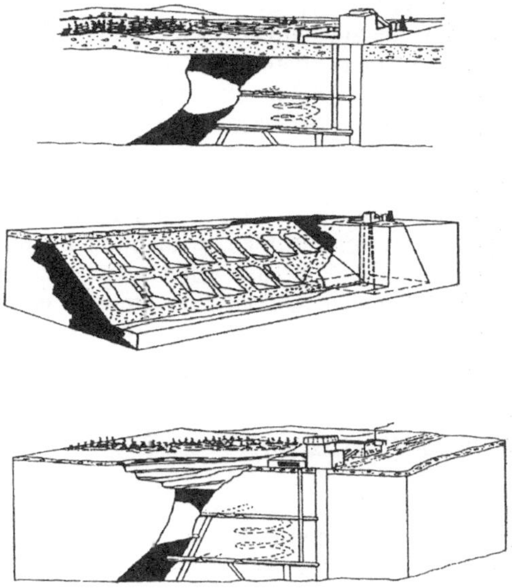

Figure 1. Common shallow stope configurations in hard rock mines

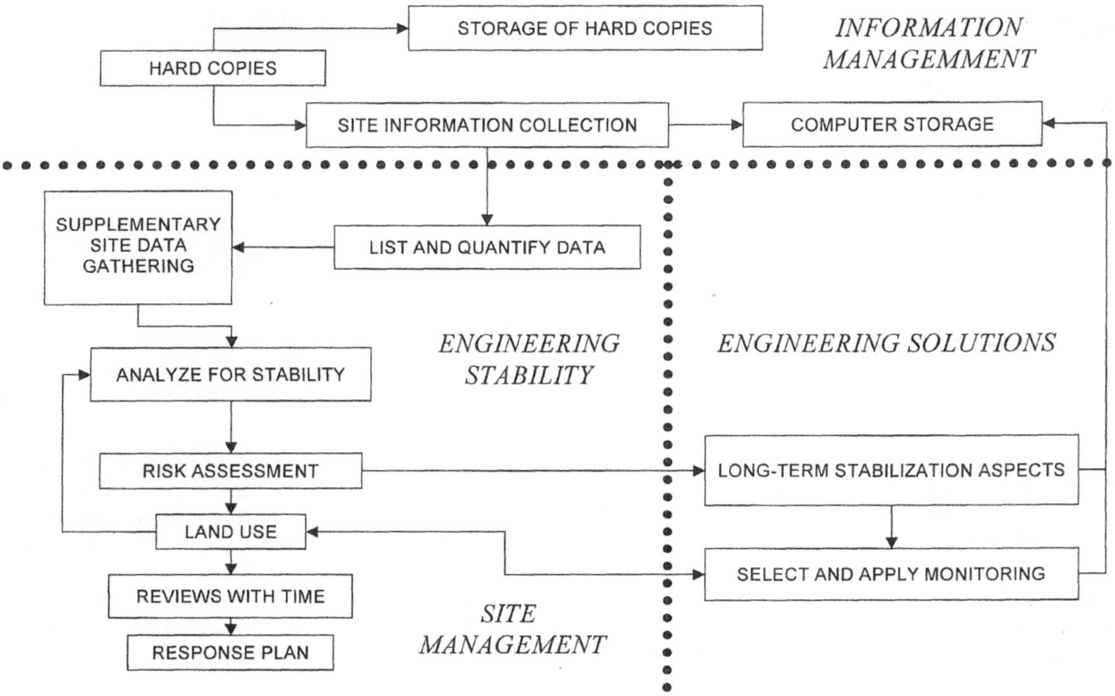

Figure 2. Shallow stope decommissioning flow chart

affecting surface integrity. However, the comprehension and range of knowledge required of responsible applicants, the general engineering consulting community, and regulatory agencies are currently insufficient to achieve the production of complete and relevant mine decommissioning plans. A mine decommissioning standard has been written for provincial and territorial governements to meet this need [Bétournay, 2001]. The information contained in the Standard includes the philosophy, experience, recommendations, as well as technical information and developments, related to the long-term, near-surface rock mass movement aspects of mine closures, on which to base best practices for rational decommissioning plans. It is presented in a step-by-step format in Figure 2, reflective of the progressive flow of the plan, with point by point prescriptives. The flow also reflects the basis and the need for complete and accurate site information, which is key to stewardship long after the site is inactive.

2. INFORMATION MANAGEMENT

Collating and evaluating a mine site's information is the first step in keeping the accurate records and data that will be necessary for supporting the development of the decommissioning approach and content as well as maintaining a foundation on which good site management can be pursued long after the mine becomes incative. Fully capturing site data is best performed using modern computing and database management technology. The original information can always serve as back-up and to re-initiate a portion or all of a database and should be archived with the regulatory agency.
The database must have an adaptable structure, provide easy, selective data retrieval, provide interface capabilities, and capture and provide data that is current and relevant. A good example of such a database was established by CANMET [Vance et al., 1995] as part of major projects under Canadian Federal-Provincial Mineral Development Agreements.

The database would contain detailed information of mine features:

1) Word-processed descriptives of mining property activity, location, ownership, shallow stope extraction, geomechanical data and ground control measures.

2) Scanned spatial information of mine features in the form of mine plans, sections, map as well as 3D graphical format of underground openings combined with overburden, surface infrastructure, bodies of water, tailings areas; where possible combined with GIS maps, borehole information, monitoring and rock mass displacement aspects, site remediation measures, permafrost-based rock mass ratings,

characterization aspects, permafrost limits and variability, bedrock weathering/rock mass variability, etc. and designation of underground aspects: location and details of filled stopes, extent of underground failure, zones of different types of ground support, rock mass quality, location of bulkheads, main geological structures, etc.

3) Important written, graphic and photo records and references, scanned

4) Results of physical stability analyses and risk analyses, scanned, and

5) Monitoring instrumentation type, location, data, scanned

3. ENGINEERING STABILITY ASPECTS

The design and ground support for shallow mine workings in permafrost terrains carried out during the life of the mine must take into consideration the provision of long-term stability and prevention of surface disruption. This becomes a critical issue when rock mass environments can be inherently difficult to stabilize when the benefits of ice-filling may disappear in the long-term with thawing. Despite the advancements in the field of rock mechanics, shallow rock masses remain complex environments for which to design. But common failure types exist. These may be precluded from occurring if the presence of ice in the rock mass in effective enough in mitigating discontinuity and rock mass disintegrating effects. However, warming of the rock mass can allow the uncontrollable cycle of ice thaw and failure. Unlike deeper stopes, where extraction takes place surrounded by rock, shallow workings are situated at or very near overburden, bodies of water or infrastructure.

A sequence of steps to identify and quantify the long-term stability of shallow mine workings is provided in Figure 3. It is formulated as a process which guides the practitioner from site reconnaissance to stability evaluation and priorization of problem identified. The end result of using this sequence will be the development of several factors of safety representing the cross-section of possible failure mechanisms. This process will be highlighted in the remainder of this section.

A detailed understanding of rock mass properties and extracting sufficient engineering characteristics about the rock mass are required to: obtain a true sense of site conditions, carry out stability analyses with a maximum level of confidence, and extract potential future dominant failure influences and types. Deposit characteristics (deposit emplacement, multiple ore zones, shape, dimensional extent and attitude, footwall and hangingwall contact conditions and folding /faulting trends) and host rock characteristics (rock types and formation history, folding and faulting trends, distribution and ice content) are to be collected and used in quantifying rock mass quality.

The most important step in providing information on rock mass environments and how a rock mass will behave is to carry out a comprehensive study of discontinuities and how in this case ice acts as a bridging/strengthening agent. Most failure types originate from control by discontinuities and rock mass fabric (bedding, foliation, etc.). This survey should be carried out for the rock mass surrounding the shallow stope and in the surface crown pillar to capture the variability and dominant structures that can influence failure.

Quantification of rock mass quality is the starting point to understand the geomechanical aspects resulting from rock and rock mass characteristics. The application of a characterization system will yield a quantitative ("Q") and qualitative value for a given rock mass that can be used in the stability analyses, compared to rock masses at other sites, and be part of a greater norm used by regulatory agencies. The NGI system [Barton et al., 1974] is adopted within the Standard because it is more widely used, is derived from a large number of case studies from shallower depths and incorporates parameters known to be influential in the mechanics of shallow stope failures. Provision to improve the Q value should be made if ice has a beneficial effect. In the case of shallow rock masses, a very poor quality rating, e.g. $Q<0.1$ represents low shear strength resistance; higher numerical values, $Q=1.0$ to 15, reflect block failures as a dominant failure mode, and very high quality e.g. $Q>25$ leads to rock mass tensile strength resistance as the dominant controlling element (while major discontinuity effects still need to be considered). It is therefore imperative that characterization of all surrounding rock mass units be performed thoroughly with a view to understand their behaviour.

The characterization results obtained can be used to compare the site conditions to common rock mass environments in which a possible number of failure mechanisms can occur, Figure 3. Failure in the context of engineering refers to the inability of a material to fulfill its design requirements, here denoting rock mass movements that start from the stope and develop to the top of bedrock. It has been shown that failures of shallow workings will not always begin within the surface crown pillar above the shallow stope [Bétournay, 1995]. It could start from the stope hangingwall or footwall. However, most of the time, the failure path runs close to or in

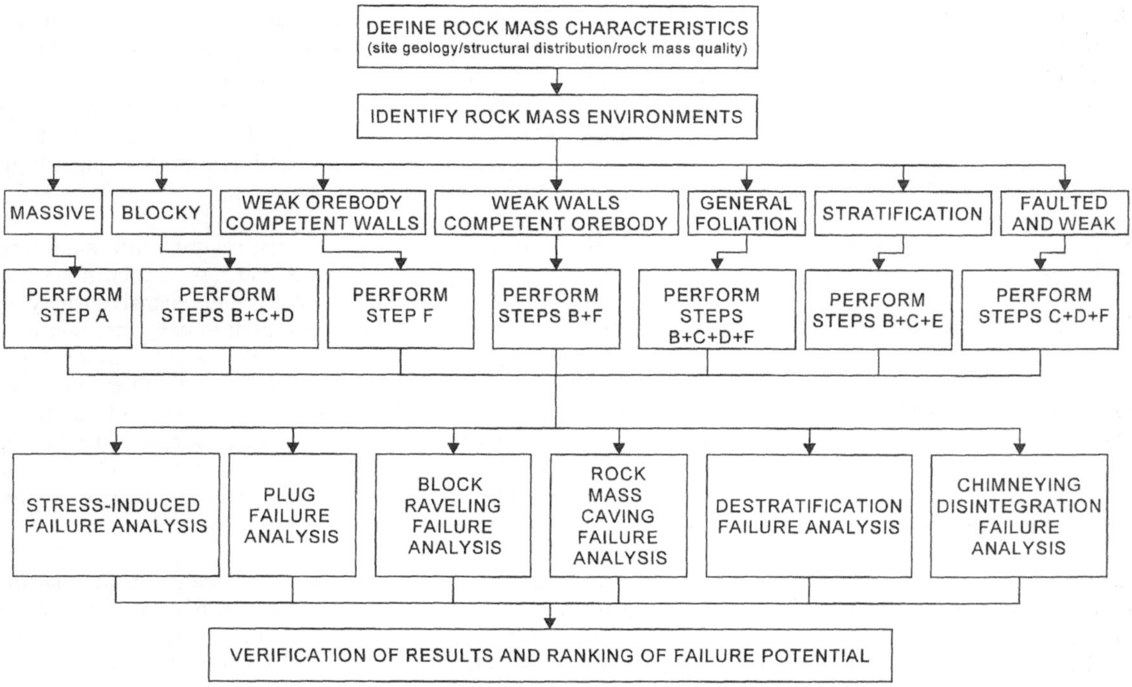

Figure 3. Shallow stope stability evaluation flow chart

the surface crown pillars. Design and evaluation of future behaviour should therefore address the question "Which portion of the rock mass is mobilized in the failure to surface?" rather than "What should be the thickness of the surface crown pillar?". The basis for analysis focused on failure mechanism/path rather than conventional approaches which rely only on the surface crown pillar, e.g. the thickness to width ratio, which can falsely lead to an assumption that a thicker rock mass is safer.

There are six types of common failure mechanisms associated with shallow stopes of hard rock mines stemming from the rock mass environments normally hosting metal bearing orebodies, Figure 3: Rock fracturing (rupture of the surface crown pillar and collapse into the shallow stope), plug failure (sudden fall of the surface crown pillar, delineated by its peripheral boundaries), raveling (block by block failure without a self-supporting cavity form reached, including the special case of delamination, the raveling failure involving sliding or buckling of thin rock layers at stope boundaries), destratification (tensile failure of stratified rock at wall contacts or along the surface crown pillar span), chimneying disintegration (the upward progression of disaggregation within a weak rock mass, by gravity, forming a cavity with limited extent, developing from an underground opening toward surface), and rock mass caving (rock mass break-up and gravity mobilization of blocks (flow) leading to a progressive failure front moving towards surface). The Standard contains the influence of rock mass properties on their occurences, common conditions for their development and cases studies profiling the basis for their occurences.

Analytical limit-equilibrium equations returning factors of safety have been derived for these failure mechanisms [Bétournay, 1995] and been outlined in the Standard with examples. These have been developed to incorporate the capability to arrive at the ultimate failure outline, for comparison to the location of bedrock and surface elements, while considering the effects of the controlling stability elements and the mechanics of failure.

Other specific stability analysis approaches are applicable. The most flexible technique is numerical modelling. It provides the possibility of obtaining approximate solutions to the behaviour of surface and/or underground excavations while considering a large number of influencing factors such as tectonic stresses, rock and rock mass properties, geometry and proximity of openings, and ground support, provided to the entire rock mass peripheral and far-field to shallow stopes.

A particular code must be selected based on a proper understanding of the case and the anticipated failure mechanism. For example, block type failures require distinct element or block code applications (in this case ice is not a factor in maintaining rock mass integrity). Weak materials with large displacements should be modelled using non-linear codes, such as elasto-plastic ones [Bétournay, 1995][Mirza et al., 1989]. In cases of competent, poorly-jointed rock masses, or integral masses, the presence of ice-filled discontinuities can signal the need for visco-elastic rather than elastic consideration of behaviour depending on the amount quantity and effective support of the ice.

As a third approach, empirical methods can establish the range of behaviours with rock mass characteristics defined for a large number of case studies. Only one dedicated empirical approach has been established to evaluate the relative stability of specific cases while considering the wide body of shallow stope experiences [CANMET Contract, 1990]. It is based on 230 shallow stope case studies, including 32 failures, in which a best fit line is derived to separate failed and unfailed cases. It does not contain data from permafrost mine failures. An expression was derived to arrive at limit surface crown pillar dimensions (Fs=1), by considering the rock mass quality Q. A process to calculate a "stable" (as opposed to "unstable") surface crown pillar is developed there to calculate the required span or thickness of the surface crown pillar only. The approach has been shown to be reliable for application in blocky rock masses where only ravelling and block caving will predominate. This is not expected in permafrost rock masses unless thawing occurs.

The advantages and disadvantages of each method are covered in detail in the Standard. A number of caveats related to the application of the methods and consideration of the results are also included.

4. SITE MANAGEMENT

Increasing concern over the impact of mining activities on public safety and the environment is focusing considerable attention to the identification of site hazards and associated risks. In a trend to strengthen regulations, coupled with new approaches to liability and responsibility, regulatory agencies are requiring that better defined and derived long-term site risk assessment and mitigation be achieved during mine decommissioning.

Geotechnical material, containing ice or not, is a good example of nature imposing variations/uncertainties to material conditions. Owing to the unpredictable distribution of rock mass elements, their variability and that it is difficult to completely quantify them, engineering such material is radically different than man-made materials such as concrete or steel. Therefore, the risk of failure is inherently greater for shallow rock masses, especially considering the much longer life span and limitation of the remediation effect compared to structures built with man-made material, designed for maximum safety and functionality with effective remediation.

To expect realistic behaviour, realistic assumptions for material properties must be made based on the consideration of all geomechanical elements, their distribution and their range of behaviour in combination to produce various outcomes. Often, if problems are understood, simple methods of analysis will provide adequate solutions. This becomes very powerful when these are based on specific material behaviour mechanisms. The approach adopted in the Standard is to perform a detailed quantitative analysis of risk at the time of decommissioning when the information can be gathered and decision about site remediation and management must be taken.

Risk is the product of the combination of the probability of occurrence of a hazard and a quantification of the possible results. A hazard with no likelihood of occurring is not a risk. A hazard with few consequences leads to a very low risk. Present day consequences may change in the future. New infrastructure or housing developments are often placed above inactive minesites. These increase the consequences of mine instabilities which may not even be known to exist.

Risk analysis [Canadian Standards Association, 1991] must be performed with sufficient scientific detail and with clarity for the results to be understood by others and to be comparable to other analyses using the same approach. Figure 4 outlines the recommended pathway for shallow stope risk analysis. The hazards associated with shallow stopes result in physical rock mass degradation that will have consequences on surface elements, near-surface buried infrastructure or acquifers. The failure mechanisms outlined are considered the hazards against risk which will be evaluated.

The standard factor of safety has been used so far as a reference point for a desired level of protection against failure. Since it is based on single input parameter values, assumptions and uncertainties are associated with the value used. A failure can occur at a factor of 2 as well as one of 5. Probability concepts describe the effect of the variability of input parameters, into limit equilibrium equations. The results are presented as a probability that a factor of safety will be less than a selected value. A probability of failure has definite bounds 0 to 1. The uncertainties associated with geomechanical stability analysis is better treated by this method. Several examples of such applications to geotechnical engineering exist [Call et al., 1982][Priest and Brown, 1983].

Considering the value of each parameter used in a limit equilibrium equation to be normally distributed leads to a quantity also normally distributed. This allows for values at each standard deviation on either side of the mean to be used in combinations, producing 2^m solutions (m = number of variables) which can be defined in a normal distribution [Rosenbleuth, 1972]. In deciding the minimum level of safety associated with a given case and anticipated failure, use is made of the quantity $P(Fs<X)$, the probability that the factor of safety will be less than a certain value (X) such as 1.0. The probability can be calculated based on the mean and standard deviation of the factor of safety values obtained from the combinations or parametric values [Ang and Tang, 1975]. The standard povides detailed calculations and examples for obtaining such values.

What is a reasonable required value against which this probability can be compared? No standard exists at the moment to indicate at what level should be adopted for a particular shallow mine working failure and its consequences. Safety factors adopted in soil and rock mechanics analyses are variable depending on the duration of the problem evaluated and its consequence of failure [Priest and Brown, 1983][Piteau and Martin, 1977]. Very long-term serviceability where public access is allowed is rated at a maximum probability of failure of <0.5%. However, no consideration in the time frame for shallow workings (perpetuity) has been derived.

The final step in the risk analysis is to evaluate the impact of a hazard on potential receptors. Ideally, this will be quantified with units that are common to different receptors. The Standard presents a cross section of hazard impact consequences that could be related to shallow stope failures and be used by various stakeholders. The categories include public aspects (safety, water source, social , cultural, trust, services), economic aspects (landowner, public, industry) and environmental aspects (species, habitat). Within each aspect, major to no impact categories are used giving a total of 44 consequences. Ordered from a pairwise comparison of all the consequences, these are allocated points which total 1,000. By adding up all of the impact consequences, the total rating value can be used against the probability of hazard occurrence to arrive at a risk value for a particular failure mechanism.

5. ENGINEERING SOLUTIONS

Engineering solutions are directed at removing or mitigating hazards identified through the stability analysis and risk assessment portions of the decommissioning plan activity. Two options are available to the site operator to control the long-term behaviour of rock masses around shallow mine workings leaving the site as is and carry out long-term monitoring and/or mitigating hazards by removing, or engineering the stabilization of potential unstable rock mass elements.

Making a site secure on closure requires consideration of hazards combined with appropriate site stabilization applications. Benign action such as doing nothing or providing surface indications of the underground stope location and warning signs or monuments (which can also act as surveying stations) posted at all location of open holes or around potential future affected areas, do not prevent access or eliminate a hazard. Before a method is selected, it is important to assess the level of risk, the associated cost and value of the mitigating method, and the future value of the mine opening.

Table 1 outlines the methods available to counter site risk. Specifications for material and installation can be found in other references [Nova Scotia Department of Natural Resources, 1995][Ontario Ministry of Northern Development and Mines, 1995].

In reviewing these options, the following aspects must be considered: use of several but complementary options (signs and others, pillar removal and backfilling, etc), land use and site restriction planning, potential for future mining activity, vandalism, positioning on competent bedrock, preventing fill run-out by using bulkheads underground, possible creation of zones of failure around the removed pillar boundaries.

Monitoring, as well as acting as a means of preventing accidents from unknown conditions, should be used to evaluate the location and progression of instabilities, supply information to verify design assumptions, stability assessments data, and ground control means, review the relevance and performance of a site remediation and management plan with time as well as the need for supplementary stabilization measures.

Monitoring is performed either visually or with the help of instruments. A basic program of recurring site visual inspections with an established frequency of visits should be established by owners, with review of information by the stakeholder community. At a minimum the following aspects should be reported on: development of surface depressions, lowering of surface water bodies, visible rock mass, fill movements, integrity of surface infrastructure, integrity of remediation measures, integrity of surveying stations, monitoring instruments.

within a length in the borehole [Dunnicliff, 1993]. While extensometers will be less expensive when a few installations are made, TDR cables are more economic when larger numbers are installed. Both have performed well in frozen rock mass applications [Charette, 1992].

Table 1. Shallow stope remediation options.

Method	Application	Risk Reduction	Effective Life	Cost
signs/monuments	warn and alert	none	temporary, <25 years	<$200 each
fence (chain link)	limit access	preventing consequence of hazard	temporary, <50 uears	$80-200 per linear meter
concrete cap	prevent access	preventing consequence of hazard	long-term, >50 years	>$4,000/m^2 of area coverage
backfilling	prevent access	preventing hazard	permanent	>$40/m^3
removal by blasting	prevent access	preventing hazrd	permanent	>$15/m^3

Every instrument installed on a site needs to be selected and placed to assist in answering specific questions as well as to respond to accuracy and economic considerations based on the site's hazard conditions and risk. By far, the greatest problem associated with instruments is the lack of long-term dedication by proponents to follow through on performing regular measurements after the plans and installations are carried out. The Standard provides details on the site condition, problem definition, instrument consideration, layout planning, instrument cost, installation, maintaining detail on specifications, and data collection, interpretation, and storage steps necessary to complete a relevant instrumentation program.

Of the available instruments, two types are recommended for long-term use. Time Domain Reflectometry, (TDR) using grouted co-axial cables in boreholes [Dowding et al., 1989][O'Connor and Dowding, 1999][Charette, 1992][Aston and Charette, 1992], can be used for general rock mass monitoring and supplemented with multiple anchor rod extensometers when movement location and type has been identified in order to provide more accuracy on the movements. TDR detects and measures the axial or transverse movement across the cable at any point along its length. It is accurate in determining the location of the displacement as well as relative movement. Extensometers, while responding only to axial deformation, provides accurate displacements

The Standard also reviews the cost of remediation and monitoring options releted to risk-based decisions.

These include total costs associated with a decision to reduce a risk (i.e. remediation plus monitoring) plus that of the consequences that can ensue given that decision. For example, supplementary to the application cost, the consequential costs associated with backfilling are zero due to the elimination of the hazard(s). In another case, to the cost associated with capping a weak rock mass (and no instrumentation) are those associated with the risk that a failure occurring beyond its boundary will have a quantifiable effect (e.g. cost of capping $800,000 to prevent a 10% failure probability and a 2 % chance that $50,000 of damage to a house occurs totals $800,000 plus 0.1x0.02x$50,000, or $800,100).

The total costs of each option can then be compared before a total site remediation plan is chosen.

6. SUMMARY AND CONCLUSIONS

Shallow stopes are inherently difficult mine openings to design for, owing to the weaker terrain that is encountered at shallow levels. This may be mitigated by the stabilizing effects of ice in rock mass discontinuities. But thawing in the long-term should be evaluated and prevented. Such ground control problems must remain the focus of long-term stability when minesites are closed. A national standard has been written to provide mine operators, regulators and engineers of record, a reference with complete and detailed coverage to elaborate a decommissioning plan that, while it provides the best available technology not at excessive economic cost, will form the basis on which solutions can be selected that will mitigate long-term consequences for the benefit of the population, the area infrastructure, as well as the environment.

7. REFERENCES

Ang, A. and Tang, W.H., "Probability concepts in engineering planning and design", John Wiley & Sons, New York, 1975, 409p.

Aston, T. and Charette, F., "Installation and monitoring of three abandoned mine crown pillar sites, Timmins, Ontario: August-October 1992", CANMET Report MRL 93-020(CL), Natural Resources, Canada, 1992, 57 p.

Barton, N., Lien, R. and Lunde, J., "Engineering classification of rock masses for the design of tunnel support", Rock Mech, 6 #4, 1974, pp. 189-236.

Bétournay, M.C., "A design philosophy for surface crown pillars of hard rock mines", CIM Bulletin, 80 903, 1986, pp. 45-61.

Bétournay, M.C., "The stability of shallow stopes of hard rock mines", McGill University Ph.D. Thesis, 1995, 611p.

Bétournay M., "A Canadian Standard for shallow stope decommissioning", CANMET Report MMSL01-005(CR), Natural Resources Canada, 2001, 180 p.

Call, R.D., Caldwell, J.A., Larson, N.B., "Optimization of open pit angles from the probability of failure", Canadian Institute of Mining and Metallurgy, 4th Open Pit Operators Conference, Edmonton, paper #22, 1982.

Canadian Standards Association, "Risk analysis requirements and guidelines", CAN/CSA-Q634-M91, 1991, 42p.

CANMET Contract 01sq 23440-8-9074, "Crown pillars stability back analysis", Golder Associates, 1990, 118p.

Charette, F., "Installation and Monitoring of three abandoned mine crown pillar sites, Cobalt, Ontario", CANMET Report MRL 92-095(CL), Natural Resources, Canada, 1992, 60 p.

Dowding, C.H., Su, M.B. and O'Connor, K., "Principles of Time Domain Reflectometry applied to measurement of rock mass deformation", Int. J. Rock Mech. Min. Sc. & Geomech. Abst., 25 (5), 1989, pp. 287-297.

Dunnicliff, J., "Geotechnical instrumentation for monitoring field performance", John Wiley & Sons, New York, 1993, 577p.

Mirza, C., Lau, K.C., Bétournay, M.C. and Crawford, A.M., "Complete recovery of surface crown pillars without removal of the overburden", Proceedings International Conference on Surface Crown Pillar Evaluation for Active and Abandoned Mines, Timmins, CANMET Special Report SP 89-5, Natural Resources Canada,1989, pp. 113-122.

Ngyen, V.U. and Chowdury, R.N., "Probabilistic study of spoil pile stability in strip coal mines- two techniques compared"; Int. J. Rock Mech. Min. Sci. & Geomech. Abstr.; 21 #6, 1984, pp. 303-312.

Nova Scotia Department of Natural Resources, "Abandoned Mines Openings Hazards and Remediation Handbook", Minerals and Energy branch Information Series 23, 1995, 90p.

O'Connor, K.M., Dowding, C.H. and Su, M.B., "Monitoring rock mass deformation using Time Domain Reflectometry", Proceedings International Conference on Surface Crown Pillar Evaluation for Active and Abandoned Mines, Timmins; CANMET Special Report SP 89-5, Natural Resources, Canada, 1989, pp. 123-132.

O'Connor, K.M. and Dowding, C.H., "Geomeasurements by pulsing TDR cables and probes", CRC Press, #0586, 1999, 424p.

Ontario Ministry of Northern Development and Mines, "Rehabilitation of Mines - guidelines for proponents", 1995.

Pine, R.J. and Thin, I.G.T., "Probabilistic risk assessment in mine pillar designs", Proceedings Innovative Mine Design for the 21st Century, Kingston, 1993, pp. 363-373.

Piteau, D.R. and Martin, D.C., "Slope stability analysis and design based on probability techniques at Cassiar mine", CIM Bulletin, 70, 1977, pp. 139-150.

Priest, S.D. and Brown, E.T., "Probabilistic stability analysis of variable rock slopes", Trans. Instn. Min. Metall., A 92, 1983, pp 1-12.

Rosenbleuth, E., "Point estimates for probability moments", Proc. Nat. Acad. Sci., U.S.A., 10, 1972, pp. 3812-3814.

Vance, J.B., Walmsley, J.R. and Bétournay, M.C., "Application of the inactive mines database", Proceedings Third Canadian Conference on Computer Application in the Mineral Industry, Montreal, 1995, pp 380-389.

Mining engineering and mine design in permafrost regions

Mining in the Arctic, Olsen, Lorentzen & Rendal (eds), © 2001 Swets & Zeitlinger, Lisse, ISBN 90 5809 177 5

Analysis of thermal regime of a deep mine airway in permafrost

S. Bandopadhyay, H. Wu & M. G. Nelson
University of Alaska Fairbanks, School of Mineral Engineering, Fairbanks, AK 99775, USA

V. Izaxon
Mining Institute of the North, Russian Academy of Sciences, Yakutsk, Russia

ABSTRACT: Large seasonal variations in the temperature of the ventilating air cause changes in the original thermal field around a mine airway, through heat and mass exchange between the air and the surrounding medium. These thermal interactions have major influence on climatic quality as well as on the stability of the mine openings. An analysis of mine thermal regime allows engineering of ventilation design layouts to provide the control of the micro-climate. This paper presents an analysis of thermal regime around a deep mine airway in permafrost. From the analysis, it can be concluded that a physical model, which is asymmetric in the Y-direction has no impact on the temperature distribution in the domain surrounding the mine airway. The results also indicate that active control of the mine thermal regime would be required with a higher priority where the rockmass material around a mine airway primarily consists of gravel.

1 INTRODUCTION

In recent years, with improvements in technology for underground mining of alluvial placer deposits, the Alaska placer mining industry has shown much interest in underground mining of frozen placers. This approach eliminates one of the main environmental issues in Alaska - water pollution caused by excessive discharge of silt into water bodies. However, mining of frozen, alluvial placer deposits by underground methods creates new problems related to the excessive convergence and the potential instability of openings in "warm", ice-rich silt and gravel strata.

Large seasonal variations in the temperature of the ventilating air cause changes in the original thermal field around a mine airway, through heat and mass exchange between the air and the surrounding medium. This is especially true in the summer, when the temperature of the air entering the mine is far above the freezing point of water, causing thawing of the ice contained in the surrounding rockmass. In the winter, in order to avoid icing up of the intake airway and to protect mine workers from cold air, the intake air is often heated, and the velocity of the heated air must be kept fairly high. These thermal interactions have major influence on climatic quality as well as on the stability of the mine openings.

The stability of frozen ground is affected by thawing, which is unusually due to the passing of warm ventilation air through the underground workings. A thermal stream is invariably established between the air passing through the mine and the surrounding rock. This leads to a gradual increase in the temperature of the frozen ground and to a decrease in its ice content.
In planning the ventilation of a mine and the methods of controlling thermal conditions, it is necessary to predict the temperature variations in the air with time as well as to determine the temperature fields around the workings.

Therefore, for planning of ventilation systems and for engineering control of the mine thermal regime in a mine in the Arctic, it is necessary to predict temperature variations in the incoming air as well as temperature variations along the length of the airway. A relationship between the deformation of the airway and the temperature can be determined based on the thermo-dynamic considerations. This relationship can be developed based on the coupled theory of both temperature and thermal stress. Thus, to evaluate the stability of a mine opening in the Arctic, it is necessary to solve a heat transfer problem generally known as the "Stephan Problem", taking into account the varying thermal-physical properties of air and the rockmass.

Because the main component of the heat balance in a mine is the heat exchange between the ventilating air and the surrounding rockmass, it is of critical importance to make an accurate estimation of the mine thermal regime. An analysis of mine thermal regime allows engineering of ventilation design layouts to provide the control of the micro-climate.

2 MODELING OF HEAT TRANSFER IN DEEP MINE AIRWAYS

The nature and intensity of heat exchange between the air and the walls of an underground mine in permafrost are functions of a number of independent variables, including the thermal properties of the permafrost and air, the velocity and distribution of the air stream, the size, slope and cross-section of the airways, the ice content and initial temperature of the frozen rockmass, the temperature and moisture content of the ventilation air, and time. The method of ventilation involving heated air, which is commonly used at present, does not correspond to the natural conditions prevailing in permafrost regions. Because the rockmass is frozen, or near the phase change temperature, a small change in the thermal regime can cause thawing or freezing of ground.

A physical model of heat transfer processes between ventilating air and the frozen rockmass around a mine opening that undergoes periodic freezing and thawing can be described as follows: temperature of the ventilating air, T_a, changes along the length of the airway, z, and with time, t, as a result of the heat and mass exchange with the frozen rockmass. The heat exchange between the ventilating air and the rockmass takes place at changing rate and periodically alternates its direction. The temperature of the rockmass, T_g, changes with time and with increasing distance from the surface of the mine opening, and as well as along the airway length (Bandopadhyay and Zhang, 1990).

The temperature regime of the rockmass, the intensity of heat conduction, and the depth of thawing and freezing can be determined based on the thermo-physical properties of rock, and its ice content, the initial characteristic temperatures, and the initial heat regime of the mine. The mathematical formulation of heat conduction problems involving phase change can be expressed by the "Stephan Equation" as follows:

$$\frac{\partial}{\partial x}(k_x\frac{\partial T}{\partial x})+\frac{\partial}{\partial y}(k_y\frac{\partial T}{\partial y})+\frac{\partial}{\partial z}(k_z\frac{\partial T}{\partial z})+a_0Q=[\rho C_p+L\delta(T-T_0)]\frac{\partial T}{\partial t}\ ; \quad (1)$$

The temperature, T, in equation (2.1) is the single value unknown function of temperature at any point, (x,y,z), at any time t; Q is the rate of heat generation per unit volume; a_o is the coefficient of heat generation; k_x , k_y, and k_z are anisotropic thermal conductivities; C_p is the specific heat; ρ is the density; L is the latent heat of phase change; and $\delta(T-T_a)$ is the Dirac Delta function:

$$\delta(y)=\begin{cases}1 & y\geq 0\\ 0 & y<0\end{cases}$$

In general, the parameters k_x, k_y, k_z, Q, ρ, and C_p in equation (1) may be functions of time, either through a dependence on local temperature or through a change in local material state or condition (thawed or frozen) with time. The coefficient of heat generation, a_o, is a constant.

In equation (1), the first set of terms accounts for thermal conduction due to temperature gradients while the second set represents the heat source (or sink) due to the ice-water phase transition.

When a mine airway is excavated in frozen ground at a considerable depth, temperature fluctuations of the air on the surface may not have any significant influence on the surrounding rockmass of the airway. Most parts of the ventilation airways for deep underground placer mines will be located at a considerable depth in the frozen ground. Therefore, the analysis of unsteady heat transfer with fixed temperatures and insulated boundary conditions is a more general situation in predicting and analyzing the temperature redistribution of the rockmass surrounding a mine airway.

2.1 *Heat transfer in deep mine airway*

A specific application of the general heat transfer equation is determined by specifying the geometry, the boundary conditions, the physical properties, and the initial conditions. In this study, the geometry of the selected mine airway shown in Figure 1 is a rectangular opening (3 m x 2 m). The problem domain selected for the analysis is also a rectangular area (11.5 m x 27 m) located at a considerable depth in the frozen ground (Figure 2).

Based on the domain defined in Figure 2, the boundary conditions for heat transfer are most often considered as the following. The initial condition is

$$T(x,y,t)\big|_{t=0}=T_0\ ; \quad (2)$$

The boundary conditions are

(1). On the inside wall, 'A-B', the convective boundary is defined as:

$$-k\frac{\partial T}{\partial y}\Big|_{y=d_1+b_1} = h(T_a - T)\,,\quad 0<x<\frac{b_2}{2}\,; \qquad (3)$$

(2). On the inside wall, 'B-C', the convective boundary is defined as:

$$-k\frac{\partial T}{\partial x}\Big|_{x=\frac{b_2}{2}} = h(T_a - T)\,,\quad d_1<y<d_1+b_1\,; \qquad (4)$$

(3). On the inside wall, 'C-D', the convective boundary is defined as:

$$k\frac{\partial T}{\partial y}\Big|_{y=d_1} = h(T_a - T)\,,\quad 0<x<\frac{b_2}{2}\,; \qquad (5)$$

(4). On the top surface, 'E-F', the fixed temperature boundary is defined as:

$$T(x,y,t)\big|_{y\to\infty} = T_g\,; \qquad (6)$$

(5). On the far right outside surface, 'F-G', the fixed temperature boundary is defined as:

$$T(x,y,t)\big|_{x\to\infty} = T_g\,; \qquad (7)$$

(6). On the bottom surface, 'G-H,' the fixed temperature boundary is defined as:

$$T(x,y,t)\big|_{y\to 0} = T_g; \qquad (8)$$

(7). On the left symmetrical surfaces, 'H-D' and 'A-E', the insulated boundary is defined as:

$$k\frac{\partial T}{\partial x}\Big|_{x=0} = 0\,; \qquad (9)$$

Because of the complex configurations and non-linear boundary conditions, a close form solution to equation (1), combined with the boundary conditions of equations (2) through (9), is not readily available. Analytical solution of the above equation presents great mathematical difficulties and consequently analytical solutions can be given for only the simplest geometrical configurations and boundary conditions.

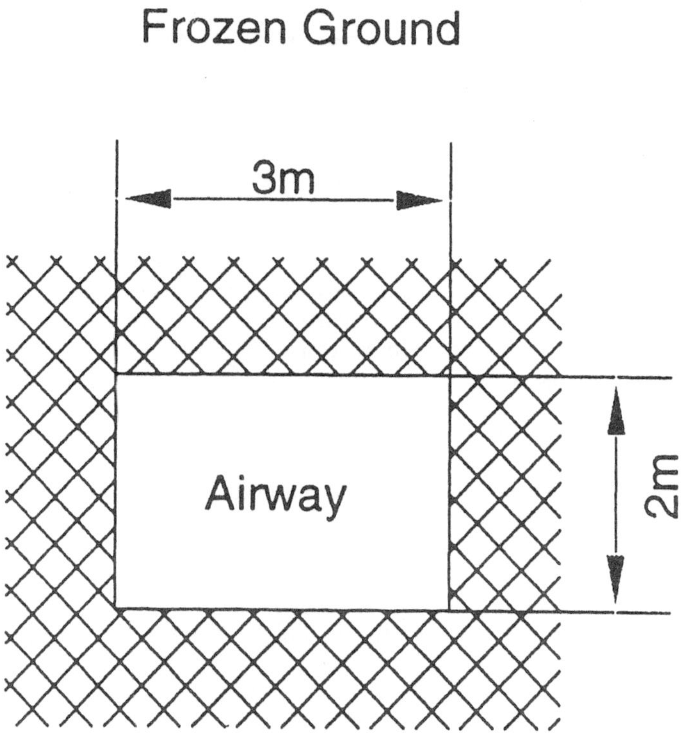

Figure 1: The geometry of the airway.

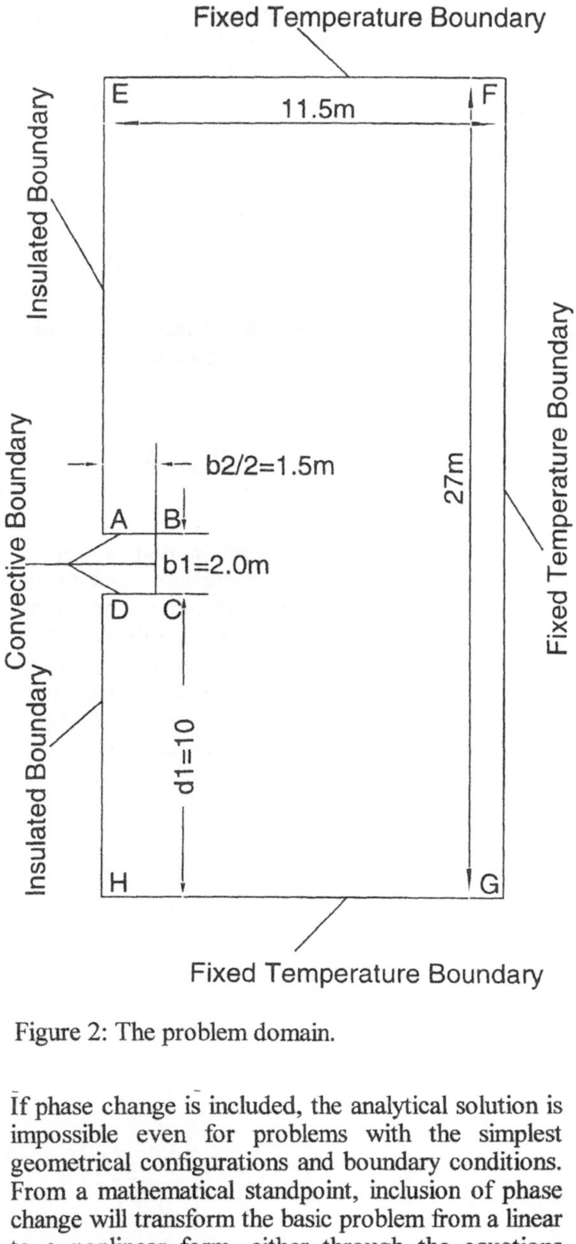

Figure 2: The problem domain.

If phase change is included, the analytical solution is impossible even for problems with the simplest geometrical configurations and boundary conditions. From a mathematical standpoint, inclusion of phase change will transform the basic problem from a linear to a nonlinear form, either through the equations themselves or through the boundary conditions. Any realistic problems thus require numerical techniques for a solution.

Numerical methods such as finite difference or finite element methods are often used to solve the problems defined by the equation (1). Bandopadhyay et al. (1995), treated the two dimensional heat transfer problem in mine ventilation with non-linear boundary conditions as two, local, one-dimensional problems with associated boundary conditions, and solved it by using finite difference approximation. Bandopadhyay et al. (1995), also presented a finite element method for solution of two-dimensional heat flow with non-linear boundary conditions, in an extensive study of temperature variations within a mine roof and the resulting stress and displacement in an underground opening in frozen ground. The finite element approach is a versatile technique and is the method selected for this study.

2.2 *Major assumptions*

In this study, several major assumptions are made during the mathematical formulation of the heat transfer problem. It is assumed that the temperature field is stationary and two-dimensional. Further, the thermal conductivity of the rockmass is assumed to be uniformly distributed and isotropic ($k=k_x=k_y=k_z$). Energy is conducted in and out of the problem domain and latent heat is released during freezing and thawing of the rockmass. Thus the problem considered here is one of conduction with a distributed energy source. The thermal properties, including thermal conductivity, k, and heat capacity, C_p, of the respective materials in the problem domain do not vary with time, nor with radial or axial direction. The physical properties such as density, ρ, moisture content, etc. of the respective materials do not vary with time and direction. The temperature, T, is the only time and location dependent variable of the total system. The system does not contain any heat sources or sinks, as in the problem analysis the changes in the mine thermal regime result only from heat exchanges between the ventilation air and the frozen rockmass. Other stationary and non-stationary heat sources are not considered here for the present analysis.

The first assumption is justified by applying the mean annual air temperature and rockmass temperature and by neglecting climatic variations. The second assumption is a mere necessity since a three dimensional model would be rather costly, in terms of programming and CPU time, and the increased accuracy would only be apparent because of other assumptions made, for example, the assumption of uniform and isotropic thermal conductivity. In general, the thermal properties of the frozen and unfrozen regions are different, but are not strong functions of temperature for each individual phase.

Under these assumptions, the reduced, two-dimensional heat conduction problem between the ventilation air and the frozen rockmass can be expressed as:

$$k\left[\frac{\partial^2 T}{\partial x^2}+\frac{\partial^2 T}{\partial y^2}\right]=\left[\rho C_p+L\delta(T-T_0)\right]\frac{\partial T}{\partial t}\ ; \qquad (10)$$

2.3 *Finite element formulation*

The continuum in a finite element analysis can be discretized by two general kinds of finite elements, triangle and quadrilateral. The linear elements in each kind of element have straight sides, while the higher order, non-linear elements, such as quadratic and cubic, can have either straight or curved sides or both. The simple elements employ first order or linear interpolation functions. The higher order elements are used to represent curved boundaries and generally have greater accuracy in their interpolation functions than that of simple elements. Because higher order elements use higher order polynomials to approximate the unknown field variables, the computational complexity involved is greatly increased. Therefore, in this research, simpler linear quadrilateral elements have been used in the two-dimensional domain, when the required accuracy can be attained. Triangular elements are used, however, for those elements which are on or near to the special boundaries of the domain.

2.4 *Finite element calculation by ABAQUS*

In the present study, the comprehensive finite element program ABAQUS was used to analyze the temperature distributions around a deep mine airway in permafrost. This program was developed by Hibbitt, Karlsson and Sorensen Inc. (1992).

To perform a finite element analysis, the parameters for solving equation (10) and those included in the initial and boundary conditions must be defined. There are numerous factors that may influence the temperature distribution and thawing of the rockmass surrounding underground openings. For a specific mine airway, it is neither feasible nor practical to include all the parameters that can be considered. Therefore, in investigating the general temperature distribution and thawing of the rockmass around a deep underground mine airway, only factors expected to have significant effects are used. The most important factors are the film coefficient and the material properties of the ground. A sensitivity analysis of the thermal regime around a mine airway was conducted by using two types of frozen soils and three heat transfer coefficients. The ranges of the parameter values used in the base cases are listed in Table 1.

Table 1: Parameter values used in the finite element calculation.

Parameter	Unit	Value
Initial ground temperature	^oC	-4
Air temperature	^oC	3
Heat transfer coefficient	$W/(m^2-{}^oC)$	17 22.72 28.4
Moisture content	%	20
Ice-water latent heat	kJ/kg	334
Fine grained soils (silt)		
Thermal conductivity (frozen)	$W/(m-{}^oC)$	2.8
Thermal conductivity (thawed)	$W/(m-{}^oC)$	1.56
Heat capacity (frozen)	$kJ/(kg-{}^oC)$	1.13
Heat capacity (thawed)	$kJ/(kg-{}^oC)$	1.15
Dry density	kg/m^3	1682
Coarse grained soil (gravel)		
Thermal conductivity (frozen)	$W/(m-{}^oC)$	4.16
Thermal conductivity (thawed)	$W/(m-{}^oC)$	3.47
Heat capacity (frozen)	$kJ/(kg-{}^oC)$	0.92
Heat capacity (thawed)	$kJ/(kg-{}^oC)$	1.13
Dry density	kg/m^3	2162.4
Dimension of selected domain	m x m	11.5 x 27
Dimension of mine airway	m x m	3 x 2
Dimension of finite element	m x m	1 x 1
Dimension of finite element on the roof	m x m	0.2 x 1

The two-dimensional domain of a general heat transfer problem in a mine airway in frozen ground, shown in Figure 3, was used in the finite element analysis.

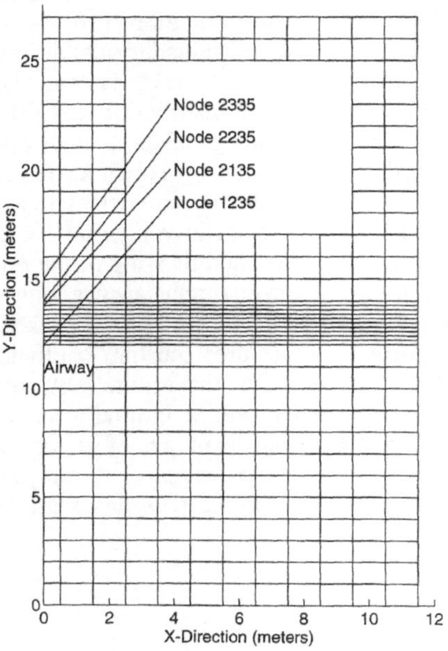

Figure 3: Finite element model of heat transfer around an underground airway for ABAQUS.

It may be noted that the domain and boundaries are structured asymmetrically in the vertical direction. This asymmetrical structure are used to select the appropriate depth of the problem domain in y-direction. By plotting the temperature contours shown in Figures 4 through 6, the vertical intersection distances from each contour line to the airway roof or floor can be determined (Table 2).

Table 2: The cases selected for an airway located at considerable depth.

Number of Case	Heat Exchange Coefficient	Material	Ventilation Time
	W/(m^2-°C)		year
1	17	silt	10
2	23	silt	10
3	28	silt	10
4	17	gravel	10
5	23	gravel	10
6	28	gravel	10

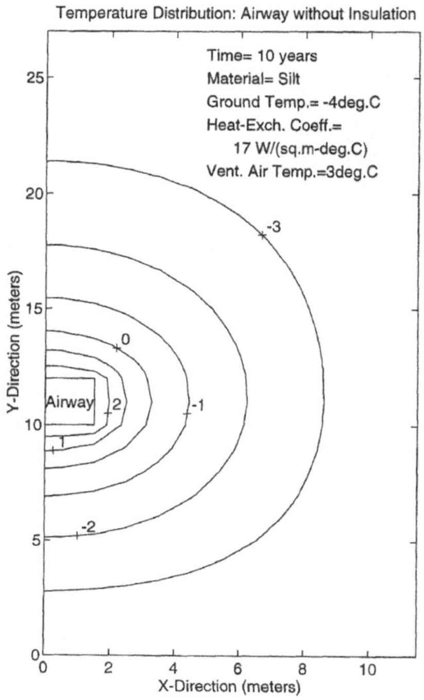

Figure 4: Temperature distribution around a deep mine airway.

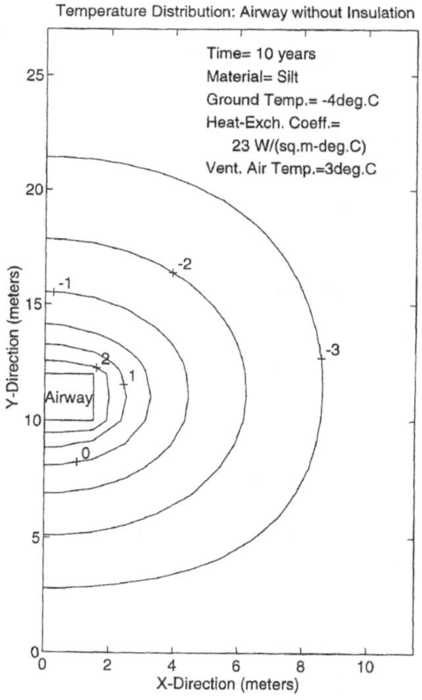

Figure 5: Temperature distribution around a deep mine airway.

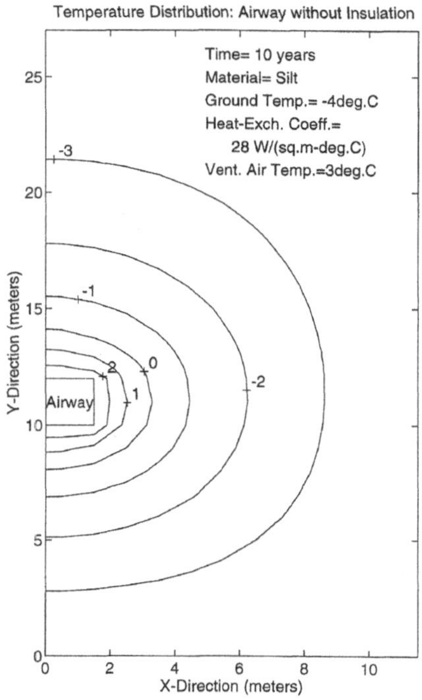

Figure 6: Temperature distribution around a deep mine airway.

Two important observations can be made from the results presented in Table 2 . First, the temperature distributions for all six cases are very similar in the region very near to the airway. The non-symmetry in y-direction in the problem domain did not influence the temperature distribution in any significant way. This is especially true when the distance is smaller than three meters. In this case, the temperature distribution can be regarded as being symmetrical to the axis parallel to x-direction, and passing through the center of the airway. Thus, the selected physical model, is acceptable even though it is non-symmetrical for analyzing the thawing region surrounding a mine airway.

The second important observation is that at the end of the 10-year period, the temperature distributions are almost the same for all cases listed in Table 3 . This implies that for all the cases investigated, the processes of heat transfer eventually reaches a steady state. The heat transfer processes beyond this time can be viewed and analyzed as a steady state process without any loss of accuracy.

Table 3: Vertical roof distances from the contour line to the airway wall.

Material	Heat Exch. Coeff.	Countour Level = -3		Countour Level = -2		Countour Level = -1		Countour Level = 0		Countour Level = 1		Countour Level = 2	
		Roof Dist.	Floor Dist.	Roof Dist.	Floor Dist.	Roof Dist.	Floor Dist.	Roof Dist.	Floor Dist.	Roof Dist.	Floor Dist.	Roof Dist.	Floor Dist.
	W/(m^2-°C)	m	m	m	m	m	m	m	m	m	m	m	m
silt	17	9.37	7.33	5.81	4.87	3.55	3.14	2.12	2.09	1.23	1.21	0.49	0.48
silt	23	9.42	7.34	5.86	4.86	3.56	3.24	2.12	2.1	1.28	1.26	0.49	0.48
silt	28	9.42	7.34	5.86	4.87	3.54	3.19	2.14	2.07	1.23	1.21	0.53	0.5
gravel	17	9.52	7.29	5.91	5.02	3.59	3.29	2.22	2.17	1.13	1.07	0.48	0.46
gravel	23	9.54	7.29	5.91	5.06	3.54	3.29	2.26	2.14	1.22	1.21	0.48	0.46
gravel	28	9.52	7.26	6.01	5.02	3.69	3.31	2.24	2.21	1.27	1.26	0.48	0.46

2.5 *Sensitivity of thermal properties*

It is well recognized that for heat conduction problems in permafrost, some thermal properties influence the thawing rate to a greater extent than others. Therefore, two types of soil (silt and gravel) with different thermal properties were used to define the frozen ground. The Stephan solution for Equation (10) depends on k, L, and C_p, since the thermal gradient is established on the basis of convective heat transfer coefficient (or film coefficient) between the mine air and the frozen ground. It is of interest therefore to conduct a parametric study of the effect of convective heat transfer coefficients on the rate of thawing.

To examine the sensitivity of these parameters, the value of the convective heat transfer coefficient was be varied, keeping the boundary temperatures constant. Several simulations were carried out and the results were plotted to study the influence of increasing film coefficient value, h, on the rate of thawing.

In the simulation results presented here, an uneven grid size was used. Elements more than two-meters away from the airway roof have larger dimension in y-direction (Δy=1 meter) than those within the two-meter region (Δy=0.2 meter). The maximum thawing depths for each element is not available at an accuracy of 0.2 meter if the thawing front is beyond the two-meter depth. A meaningful conclusion that can be drawn, however, is that during the remaining 10 year period, the thawing front for each case does not reach a three-meter depth in the vertical direction along the roof.

It is clear from Figure 7 and 8 that convective heat exchange coefficient between the ventilating air and the rockmass has significant influence on the rate of thawing. However, for different types of frozen rockmass, the degree of influence is varied. For example, when the convective heat exchange coefficient of silt was changed from 17 W/(m^2-°C) to 23 W/(m^2-°C), the time needed to thaw a two-meter depth into the roof was reduced from 9.31 years to 8.44 years. When the convective heat exchange coefficient was changed to 28 W/(m^2-°C), the time needed to thaw a two-meter roof was 8.42 years, almost the same as that for a heat transfer coefficient of 23 W/(m^2-°C).

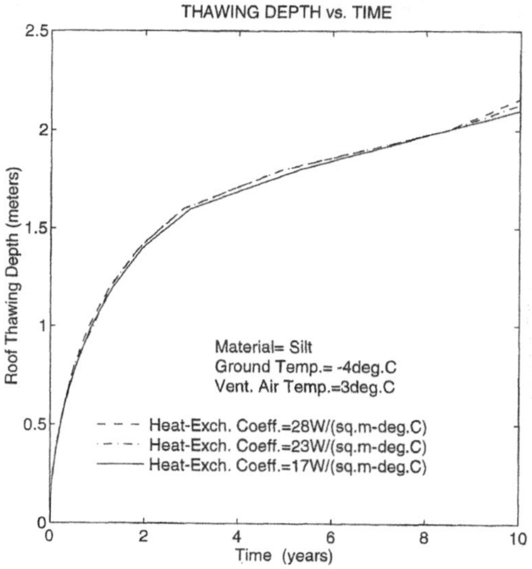

Figure 7: Thawing depth in a deep underground placer mine (silt).

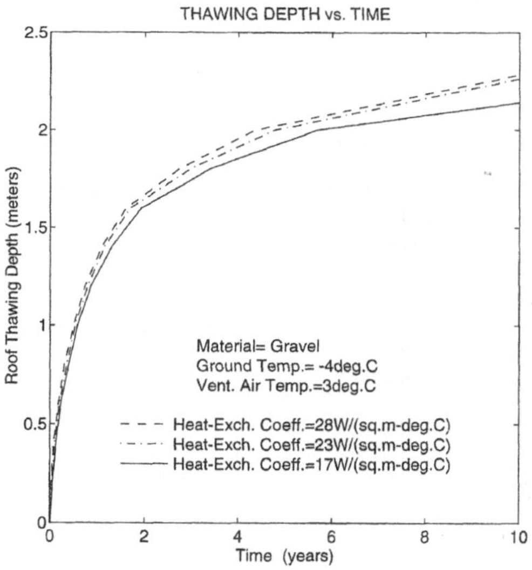

Figure 8: Thawing depth in a deep underground placer mine (gravel).

For gravel, the values of convective heat exchange coefficient appeared to have little significance. The times needed to thaw a two-meter roof were clearly reduced in comparison with those of silt under same conditions, because the heat conductivity of gravel is larger. Otherwise, the time to thaw a two-meter roof decreased significantly with increasing values of convective heat exchange coefficient. For the three film coefficients of 17, 23, and 28 W/(m^2-°C), the times needed to thaw a two-meter roof into the frozen gravel were 5.67, 4.77, and 4.43 years respectively. The time intervals for each combination of two frozen soils and three convective heat exchange coefficients for a two-meter thawing depth are presented in Table 4. The heat transfer effect tends to increase with higher heat conductive coefficient values. It is also apparent that the conductivity of frozen soil plays a major role in transferring heat.

It is also of interest to compare relative thawing depths for the two material types: silt and gravel. Figure 9 shows how each selected value of convective heat transfer coefficient influences impacts the thawing depth. A number of simulations were carried out in which calculations were extended to 10 years. Plotted in Figures 4 through 6 are the temperature profiles surrounding the mine airway in silt for various convective heat transfer coefficients. It can be seen from these figures that the temperature of the rockmass surrounding the airway is above the phase change temperature, which would result in thawing of the ground. Figures 10 through 12 show the roof temperature variations. The incremental effect tends to decrease with higher heat conductive coefficient values, over 10 years, again for mine openings in both silt and gravel. The time to thaw a two-meter depth is apparently shorter for gravel than that is observed in silt (Table 4 and Figure 9). This indicates that active control of the mine thermal regime would be required with a higher priority where the rockmass material around a mine airway primarily consists of gravel.

From these results, it can be concluded that a physical model, which is asymmetrical in the y-direction as shown in Figures 7 and 8, has no impact on the temperature distributions in the domain surrounding the mine airway. After a sufficient time, the temperature distribution in the surrounding rockmass of an underground placer mine airway tends to reach a steady state condition. The larger the convective heat exchange coefficient of frozen soil, the faster is the rate of thawing. The thawing rate is also higher in frozen gravel than in frozen silt. This is because of the higher thermal conductivity of gravel, and the smaller volume of ice-bonding material in the rock matrix.

Table 4: The time to thaw a 2-meter depth in the mine.

Heat Exchange Coeff.	Material	Thawing Depth Roof Distance	Time
W/(m^2-°C)		m	year
17	Silt	2	9.31
17	Gravel	2	5.67
23	Silt	2	8.44
23	Gravel	2	4.77
28	Silt	2	8.42
28	Gravel	2	4.35

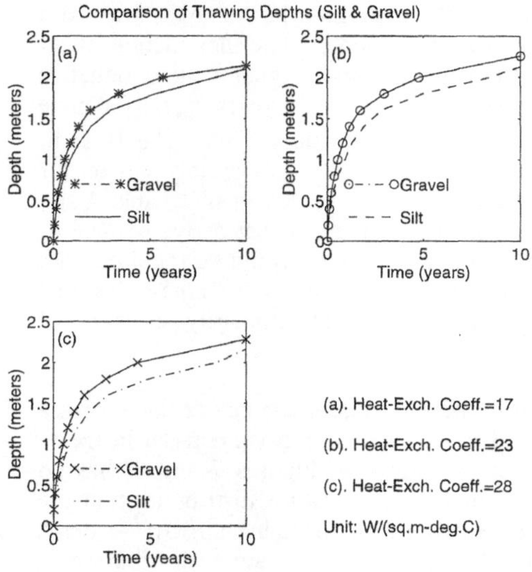

Figure 9: Airway thawing depth in an underground palcer mine.

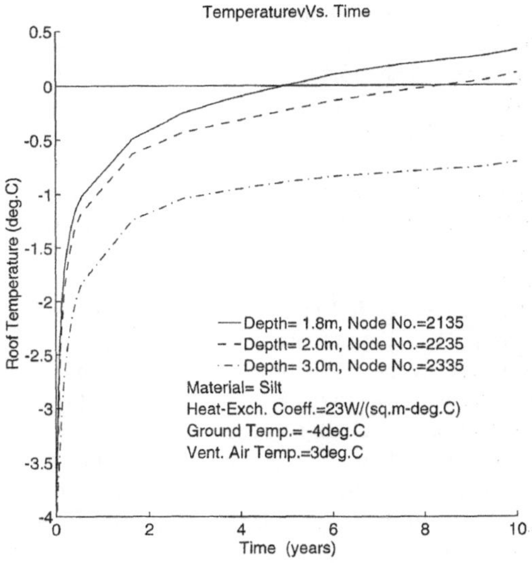

Figure 11: Temperature variations at selected roof points

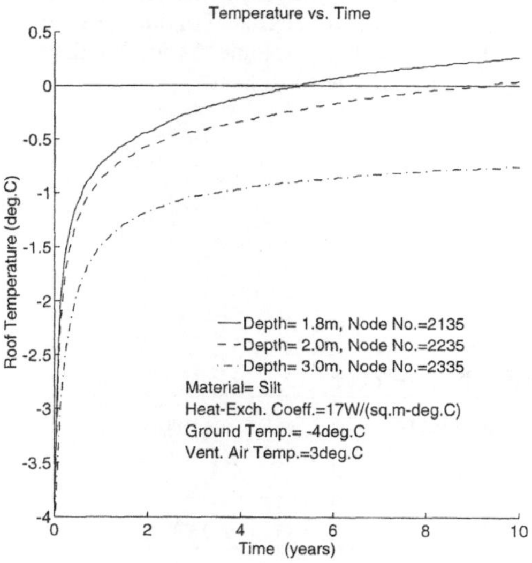

Figure 10: Temperature variations at selected roof points.

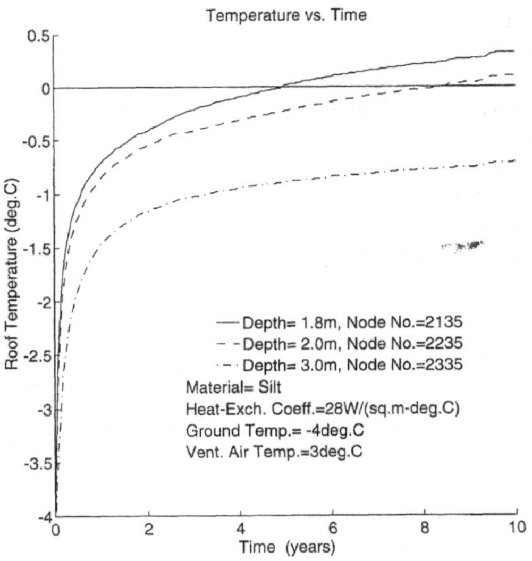

Figure 12: Temperature variations at selected roof points.

References

Bandopadhyay, S. and Y. Zhang, "Influence of Ventilating Air Fluctuations on Stress Distribution Around an Airway in Frozen Ground," transacytions, SME/AIME, Vol. 288, pp. 1819-1825,1990.

Bandopadhyay, S., K. Biswas, and M.G. Nelson,"Evaluation of a Roof Bolt Support System in Arctic Palacer Mines", Transactions, SME/AIME, vol. 296, 1995

Bandopadhyay, S., H. Wu, M. G. Nelson, and V. Izaxon, "Design of Ventilation Systems for Underground Placer Mines in the Arctic: Analysis of Mine Thermal Regime," Proceedings of the 7th U.S. Mine Ventilation Symposium, A. Wala (Editor), SME/AIME, pp.27-32, 1995.

Mining in the Arctic, Olsen, Lorentzen & Rendal (eds), © 2001 Swets & Zeitlinger, Lisse, ISBN 90 5809 177 5

Tailings deposition and dike construction at Nanisivik Mine, Nunavut

J. W. Cassie
BGC Engineering Inc., Calgary, Alberta, Canada

K. G. LeDrew
Nanisivik Mine, a division of CanZinco Ltd., Nanisivik, Nunavut, Canada

ABSTRACT: This paper presents an overview of the tailings disposal methodology and tailings retention dike design used at Nanisivik Mine, situated near the northern end of Baffin Island. Due to the extraction of potable water from one of the lakes adjacent to the tailings disposal area, the tailings disposal area was divided into an upper Surface Cell for solids retention and a lower Reservoir for the storage of supernatant water. A frozen-shale, upstream-constructed dike is used to retain the solids and to separate the Surface Cell from the Reservoir. The paper reviews the construction methodology used for this novel retention dike and the dike performance to-date. It also reviews a seepage event that occurred in 1999 and the lessons learned from that event.

1 INTRODUCTION

1.1 *Location and site description*

Nanisivik Mine is located adjacent to Strathcona Sound, near the northern end of Baffin Island, at approximately 73°03' North. The tailings disposal area is centred on West Twin Lake, which is situated approximately 6 km inland from the coast. The West Twin disposal area (WTDA), as it is known, is situated in a topographic low just east of Mt. Fuji, which is the highest landmark in the immediate vicinity (crest elevation higher than 650 m), as shown in Figure 1.

West Twin Lake is mirrored by East Twin Lake, where potable water is derived for the mine. In order to prevent possible hydraulic connection between the two lakes, the level of West Twin Lake is kept lower than that of East Twin Lake. Due to this requirement, West Twin Lake was divided into two sections, which allows the water level in the lake portion to remain below the level of East Twin Lake.

The disposal area is divided into an upper Surface Cell, for solids retention, and a lower Reservoir, where excess supernatant water is stored before it is pumped back to the mill. A frozen, shale embankment, referred to as the West Twin dike, is used to separate the frozen tailings solids from the lower water pond. In addition to tailings discharge water, the dike retains surface runoff from a portion of the hillside directly to the north and a portion of the runoff coming from Mt. Fuji. Generally each year, the retaining dike is raised by approximately 2 m using frozen weathered shale as the construction material. Currently, the dike is approximately 17 m high.

1.2 *Physical conditions*

Bedrock units exposed near the disposal area include the Adams Sound Formation quartzite, which forms a plateau to the south, the Society Cliffs Formation dolostone immediately to the north and south and the Victor Bay shale to the east and west. The dolostone and the overlying shale belong to the carbonate shelf sequence of the Bylot Supergroup, which has been intruded by mafic dikes. Both the dolostone and the quartzite are recrystallised, forming dense, massive units. Metamorphic effects are minimal, well below the greenschist facies; hydrothermal alteration is confined to intrusive dike margins. The bedrock surface is mantled with glacially derived overburden.

Climatic data has been collected since 1976 at the Nanisivik Airport, which is located approximately 10 km south of the disposal area and approximately 250 m higher in elevation. The mean annual air temperature is −15.2°C. The mean annual precipitation totals 240 mm, with on average, more than 41 mm falling in each of the months of August and September. The mean annual lake evaporation value, as measured at site by Federal government researchers, amounts to 200 mm.

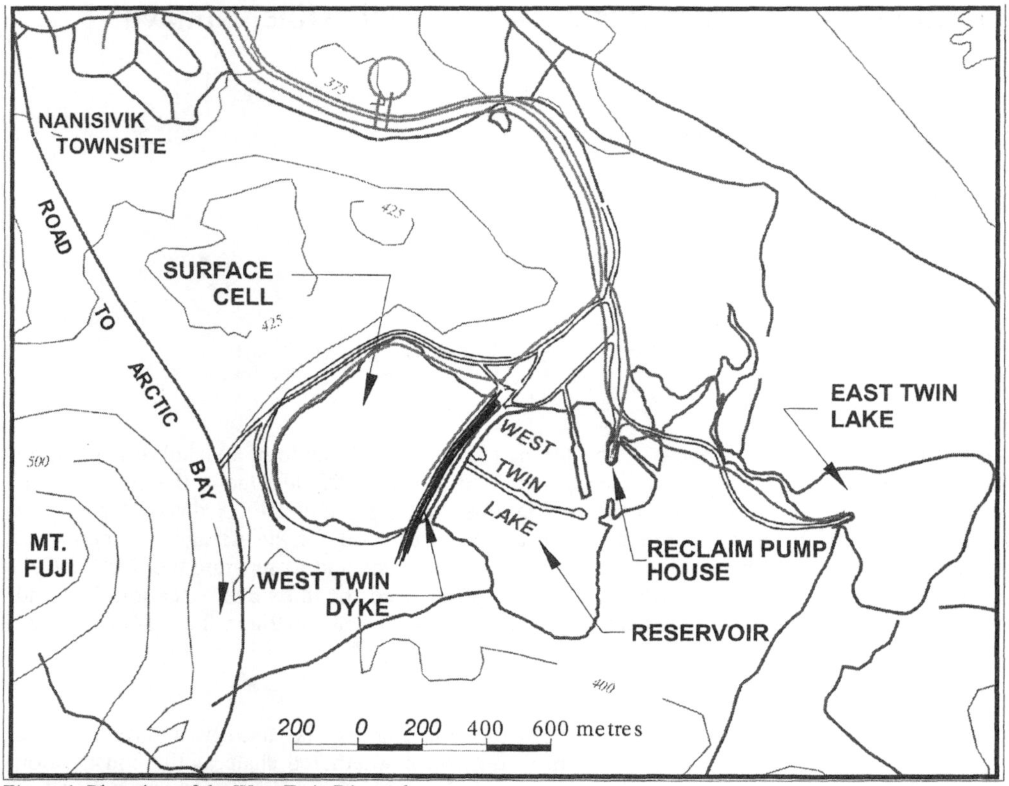

Figure 1. Plan view of the West Twin Disposal area.

IPA (1997) maps indicate that the northern end of Baffin Island is situated on continuous permafrost, with a potential for high amounts (>20%) of ground ice. One measured value from this reference notes that a mean annual ground temperature value of –12.2°C was recorded at the depth of zero annual amplitude at the north end of Baffin Island. Tordon (1997) notes that permafrost was previously found to be deeper than 430 m, as drilled from the mine workings.

Watts et. al. (1973) noted that subsurface rock temperature measurements, collected during the project feasibility study, ranged from -11.7° to -9.4°C.

2 WTDA BACKGROUND

2.1 *History of the WTDA*

The Nanisivik deposit was initially discovered in 1956 and mining began at the site in August, 1976. When the mine initially began, the proposed mine life was 12 years. B.C. Research (1975) undertook some environmental survey work in mid-1974 and no fish population was observed within the Twin Lakes system. Late in 1976, a decision was made to deposit the tailings within West Twin Lake.

The original ore reserves were 6.5 million tonnes resulting in 5 million tonnes of tailings over the initially specified mine life. But by mid-1989, 6 million tonnes of tailings had been placed in West Twin Lake and 2.6 million tonnes of proven ore reserves were still remaining. It became apparent that the mine tailings would exceed the original lake volume and that additional storage capacity would be required.

During the placement of tailings within West Twin Lake, two "causeways" developed from the beaching of tailings solids. Terratech (1989) undertook an initial review of potential options for increasing the storage potential of West Twin Lake. Five potential containment dike alternatives were provided; two of which used these existing tailings causeways. Two boreholes were drilled into the tailings causeways in June 1989 and frozen conditions were encountered to a depth of 12 to 13 m.

It was therefore decided to divide West Twin Lake into two sections using an earthen dike, situated on top of the frozen tailings causeways. The eastern portion of the lake, the Reservoir, would retain its original lake level and include one deep pocket of water for emergency tailings deposition during mid-winter conditions. The western part of the lake, the Surface Cell, would become the main deposition area for the tailings and would rise above the original West Twin Lake level. Initially, it was proposed to place tailings in the lake to a maximum height of 5 to 6 metres (Elevation 376 m) above the original lake level.

The dike was placed along a pre-existing north-south tailings causeway, which prior to being used as the foundation for the dike, extended to within 50 to 75 m of the south shoreline of the lake. Based on subsurface temperature readings within the causeway, the 0°C isotherm was at least 10 m below the original lake level (Terratech 1990). Thus, a frozen foundation for the dike was in place for a majority of its length.

On-land disposal of tailings commenced on August 27, 1990 in the northern portion of the Surface Cell. In the years since the commencement of on-land disposal, it has been necessary to raise the dike structure by approximately 2 meters annually (1994 and 1999 were the exceptions). At present, the West Twin dike is approximately 900 meters in length and has an upper crest elevation of just less than 388 meters, approximately 17 meters above the lower water reservoir elevation of 371 meters.

2.2 *Operation of the WTDA*

The upper Surface Cell and the lower Reservoir are separated by the West Twin dike, constructed of compacted shale and settled tailing. A siphon system acts as the decant line removing ponded water from the Surface Cell. The Reservoir and polishing pond are separated by a causeway and stop log system and the polishing pond water level is controlled by a concrete decant structure equipped with stop-logs.

Tailings are deposited into the Surface Cell during the winter along the perimeter to promote stratification of sediments and to clarify waters returning to the decant/siphon well area. Advancement of successive discharge points is determined by the overall height of the deposited tailings and the ability to return clear water to the decant well. Most of the excess supernatant pond water collects in a depression situated upstream from the south end of the dike. Excess water is then siphoned into the Reservoir from where it is reclaimed for use in the mill. Occasionally, the Surface Cell pond can get too shallow and tailings discharge occurs directly into the Reservoir.

Due to the extraction of potable water from East Twin Lake, it is a requirement of the water licence that the West Twin Reservoir elevation remain lower than the water level elevation within East Twin Lake.

Excess water from the WTDA is discharged to the environment between July and September of each year, with annual quantities typically ranging from 200,000 to 400,000 m^3.

2.3 *Conceptual closure of the WTDA*

When the mine is closed in the future and reclamation of the Surface Cell is undertaken, an appropriately designed thickness of shale will be placed over the frozen tailings. The placement of this layer will move the active layer upward and prevent thaw of the underlying tailings. In addition, a surface course of sand, gravel and cobbles will be placed to mitigate surface erosion. The cover layers will be graded down towards the south. A spillway will be required at the south end of the dike to divert excess surface drainage away from the dike structure and allow the surface water to flow to the downstream reservoir.

3 WEST TWIN DIKE

3.1 *Dike construction and configuration*

The first lift of the shale dike was built in 1991 with a top width of approximately 4 m, at a nominal elevation of 374 m. Since then, the dike has been raised annually, except in 1994 and 1999.

The dike has been raised in an upstream manner where each annual lift construction begins on top of beached tailings material deposited in the previous year's construction and sealing phase, as shown in Figure 2. Prior to placement of any fill, surface snow and ice is removed from the tailing platform and shale shell. During the late Winter, weathered shale rockfill is hauled and dumped, spread with a dozer, wet down with a water truck and compacted (with construction traffic) in 30 to 40 cm lift thicknesses to permit freezing and nominal compaction to take place. The downstream face of the shale rockfill is placed at approximately the angle of repose, approximately 1.5H:1V to 2H:1V, and a setback is provided on each lift.

After a lift has reached the nominal thickness of 2 m and is frozen, sealing of the upstream face is carried out using tailings discharged during the following Summer. Tailings are discharged approximately 10 to 15 m upstream of the face of the dike. The fine-grained tailings material (deposited on the reverse beach formed against the upstream face) provides a partial barrier to water migration through the shale fill and a base for future dike construction. This reverse beach is extended to a height of no more than 1 m below the top of the shale dike. During the following Fall, a small thickness of saturated shale is placed over the previous Summer's beach material in order to level out the surface for the following Winter construction. In addition, the placement of this levelling lift aids in the reduction of dusting problems.

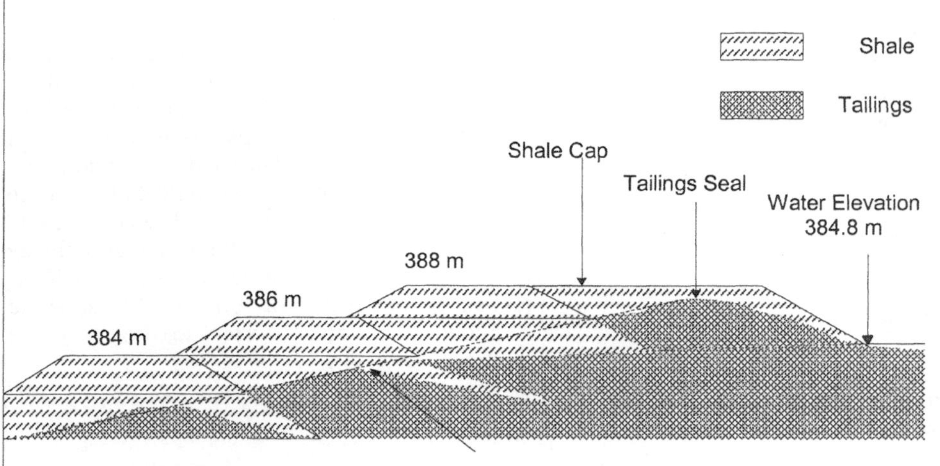

Figure 2. Cross-Section of the West Twin Dike.

An overall downstream slope of 3.7H:1V has been achieved using a nominal lift thickness of 2 m and a setback width of 4 m. Currently, the dike crest elevation varies from 387.6 m to 387.8 m. Each individual lift is referenced by its nominal top elevation (e.g. the "388" lift). In 2000, backfilling of the terraced downstream face began to form a uniform slope.

3.2 *Materials and foundation conditions*

Geotechnical information for the dike has been obtained from approximately 38 boreholes, which range from 5 to 24 m depth. The simplified stratigraphy consists of shale rockfill for the dike, tailings sand and silt, lakebed sediments and/or till, and finally bedrock (either dolostone or shale). From south to north, the bedrock surface elevation dips from approximately 369 m to a low of 361 m. The bedrock surface then rises to elevation 374 at the north end of the dike. Although the bedrock surface is somewhat irregular, it more or less reflects the expected valley geometry.

Overlying the bedrock is a combined unit of lakebed sediments (silt and sand) and/or glacial till (silt, sand and gravel fragments) that varies from 1 to 4 m in thickness. One exception to this is at the north end of the dike where no lakebed sediments/till were encountered. This location appears to be higher than the traditional lake level. Overall, the lakebed sediments/till upper surface generally follows the bedrock topography.

The tailing thickness varies from 3 to 7 m underneath the main portion of the dike and thins out towards the north end. Overlying the tailings is the shale rockfill, which acts as a frozen retention dike, up against which the tailings are initially placed.

The tailings consist primarily of a silty sand with a trace of clay but locally grades to a clayey silt. Based on 25 moisture content tests previously carried out on tailings samples, an average value of 18% was determined. Five tailings sample collected during the August, 1999 program returned moisture content values from 10% to 17%.

The shale rockfill comes from various sources near the dike and consists of angular fragments of cobble to gravel size with smaller amounts of sand and silt fractions. Based on approximately 120 compaction test results from 1993 to 1998 (Terratech 1993a and Nanisivik Mines 1993 and 1994), the moisture content varied from 10 to 40% with an average of 24%. The shale material exhibit some breakdown under abrasion and freeze/thaw conditions.

In reviewing the 38 available borehole logs adjacent to the West Twin dike, only six references to ice lenses were encountered. Given the variation in drilling methodologies used for the various boreholes, it is possible that some ice lenses may have not been recovered or observed. But, given that 38 boreholes were drilled, and only a limited number of ice lenses were observed, it is likely that there are no major, continuous ice lenses within the foundation of the dike.

The tailings are mostly (80%) composed of pyrite, which has a specific gravity of 5.0. Most of the remainder (18%) is dolomite, which has a specific gravity of 2.85. Hence, it is estimated that the specific gravity for the tailings is approximately 4.6.

Generally, for beached tailings deposits, a void ratio of approximately 0.9 to 1.1 can be expected. Based on these assumed parameters, the expected dry density for the tailings would be approximately 2,300 kg/m^3, which is approximately the value that the mine uses for planning of tailings storage. For these same parameters, the saturated moisture content would be approximately 22%. Based on the data collected to-date, it appears that the in situ moisture content of the tailings sand is just lower than the saturation moisture content. Hence, no excess ice is anticipated, a condition confirmed by the lack of ice lenses in the drilling records.

Lakebed sediments, composed of silt with some sand, are identified by their red colour. No available lab data is available for these materials. Glacial till comprises a mixture of clayey silt with some sands and trace of gravel fragments. The bedrock surface consists of dolostone or shale. Based on the mapping of surface outcrops, the dolostone is slightly weathered, medium strong and fractured, while the shale is weak and highly fractured with oxidised joint surfaces.

3.3 *Geothermal regime*

Numerous frost gauges and thermocouple strings have been installed in the West Twin dike to monitor both the active layer depth and the subsurface temperature regime.

Two frost gauges located within the dike shale rockfill show a maximum depth of thaw varying from 1.2 to 1.5 m below surface. Generally, the thawing process starts in June and ends October. The amount of active layer thaw has been decreasing steadily since readings were initiated in 1993. This observation is probably related to an increasing level of saturation within the shale fill. The increased moisture content reduces the depth of penetration of the thaw front.

There are approximately 20 thermocouple strings currently in operation on the dike. Generally, these are read once per month and the data is entered into a spreadsheet for review by mine staff.

A time lag of a few months exists between the maximum freezing (or thawing) air conditions on the surface and the coldest (or warmest) conditions measured in the ground. Instrumentation data indicates that the coldest subsurface temperatures exist typically in April and the warmest temperatures typically exist in October.

As such, the subsurface temperatures were averaged for the period 1993 to 1998 (or 1999) for the months of April and October. Generally, all the temperature contours are below zero (not including the active layer) and show a relatively consistent thermal regime within the dike foundation for a given time of year. These temperature contours rise slightly proximal to Thermocouple #13 indicating that the cold temperatures do not progress as deep into the foundation zone in this area. This data indicates that this area of the dike foundation is warmer than typical, which is likely caused by the deep-water pond which has always been situated just upstream from Thermocouples #12 and #13. Since the pond is never completely drawn down and the tailings exposed, the warm water heat source may be keeping this part of the dike foundation warmer than other parts.

Thermocouple #21 is situated near the northern end of the Surface Cell and provides subsurface readings within the tailings deposit from elevation 385 down to 376 m. Below a depth of 3 m, the subsurface temperatures are below freezing and generally colder than -4°C. At a depth of 9 m, the temperature varied from -6.9°C to -7.9°C within one year. Hence, the depth of zero annual amplitude appears to be greater than 9 m within the tailings deposit. This thermocouple also confirms that the tailings deposit upstream of the dike is frozen to a depth of at least 9 m. It should be noted that no deep water pond was ever formed adjacent to Thermocouple #21.

In summary, the data provided by the array of thermocouples appears to confirm that the entire foundation zone beneath the West Twin dike is below zero. In addition, the tailings deposit in the Surface Cell appears to be below zero. Only the tailings proximal to the deep hole near the south end of the dike may be above zero. Some investigative work is planned for this aspect in 2001.

4 STABILITY ASSESSMENT

McRoberts and Morgenstern (1974) noted that for stability analysis of ice-poor frozen soils, the long term strength can be approximated by using the unfrozen frictional strength. Since drilling information collected to-date does not indicate any significant and/or continuous ice lenses, it was assumed that the foundation materials were ice-poor. Hence, conventional slope stability analysis of the current dike was undertaken. Since the dike and tailings are frozen, stability analyses were undertaken for the case of a dry, frozen slope only (i.e., no pore pressures involved).

The slope stability software used Janbu's non-circular method of analysis in the commercially available program SLOPE/W, under license from Geo-Slope International. The stability model was configured such that the failure surface search was forced into the materials below the shale dike envelope to evaluate deep seated failures through the foundation materials.

The analyses were performed using the pore water pressure ratio, r_u, equal to zero to simulate a dry, frozen slope. Additionally, a 0.05g horizontal seismic load (seismic building code zonation) was applied to simulate earthquake conditions. Frictional angle values (ϕ') of 23° to 33° were assumed for the tailings sand.

Based on these parameters and for the lowest tailings value of 23°, the Factor of Safety ranged from 1.6 to 1.8 for various foundation configurations under static loading and from 1.3 to 1.5 for seismic conditions. All of these values are in excess of current Canadian Dam Safety Guideline values. Laboratory testing has been recommended to confirm the frictional value of the tailings.

Thermocouple information indicates that the footprint area below the dike is frozen, generally down to the bedrock surface. The dike, consisting of both the shale dike material and the underlying tailings foundation material, when frozen, act as the structural element to retain tailings solids and water, before frost penetration occurs during the following winter period. Therefore, if the entire dike block, consisting of shale and tailings, was proven to be frozen down to the bedrock surface, then it would be valid to model the two materials as a "rigid block". This block, similar to a rigid structure such as a concrete buttress dam, would retain any possible thawed zones of tailings. This zone could apply both water and soil pressures onto the upstream side of the dike.

A rigid block model was formulated to assess the factor of safety for sliding of the dike along the foundation materials. An assumption was made that the foundation material was the shale bedrock, which is weaker than the dolostone unit, which also underlies a portion of the dike. The analyses were conducted by assuming that the shale dike and underlying foundation tailings were completely frozen down to bedrock and acted as a rigid block. It was further assumed that a thawed zone exists directly behind this rigid (frozen) block and that the thawed zone applies both hydrostatic and active earth pressures onto the rigid block.

When the frictional strength of the foundation (ϕ_f') is 20°, a Factor of Safety of 1.95 is determined for the case where both the active earth pressure and water pressure act against the rigid block. Similarly, the analysis for the case where the foundation soil strength is ϕ_f'=15°, the lowest calculated factor of safety is 1.43. Therefore, from these analyses, all factors of safety against sliding for the rigid block are greater than 1.43.

The data reviewed indicates that no significant massive ice lenses or bodies are situated within the tailings. Hence, it is unlikely that significant creep deformation of the dike will occur.

5 1999 SEEPAGE EVENT

On June 21, 1999, as the tailings seal was being discharged onto the upstream side of the 388 Lift, a small amount of seepage was observed discharging from the 376 Lift, near Thermocouple #13 situated near the south end of the dike. On July 5, 1999, seepage was noted discharging on the 376 Lift at both the locations of Thermocouple # 13 and Thermocouple # 18, near the north abutment of the dike. Estimated seepage quantities at that time were 5 l/min. at Thermocouple # 13 and 2 l/min. at Thermocouple #18; the seepage water discharging was clear of sediment. At this time, the elevation of the tailings discharge line had been allowed to progress to a level less than one meter from the crest level, which was in contravention of the recommended practice for beaching.

After the seepage was observed, a review of thermocouple data for Thermocouples #13 and #18 was undertaken. At Thermocouple #13, the temperatures in June 1999 ranged from about –5° at 3.5 m depth to about –6° at 9.5 m depth. These temperatures were similar to those measured in June 1998. In July 1999, the temperature at 3.5 m depth indicated slight cooling but the bottom portion of the string was warming significantly; -5° at 7.5 m depth and –3.5° at 9.5 m depth. Although no readings were available for July 1998, interpolation of the 1998 data sets indicated that a temperature of about –5.3° would have typically been expected at the 9.5 m depth. Hence, it appeared that some atypical warming was occurring at depth.

At Thermocouple #18, the subsurface temperature profile for June 1999, was colder than –8° for the top 4 m, warming to –4.5° at 6 m depth. This profile was different than from June 1998 where at 6 m depth, the reading was –9.5°. Hence, at 6 m depth, the 1999 reading was about 5° warmer than 1998. The July 1999 readings also exhibited a continuing warming trend below a depth of 4 m; at 6 m depth, the temperature was about –2°. Hence, the thermocouple data indicated that atypical warming was occurring at a depth of approximately 4 to 6 m. This depth is generally located at the bottom of the frozen shale dike fill. The dramatic warming in the subsurface thermal regime was best explained by convective heat transport caused by flowing water.

By July 9, 1999, the seepage discharge at Thermocouple #13 was occurring over a 35 m wide area, but the seepage at Thermocouple #18 was reduced to a trickle. Mine staff observed that the Thermocouple #13 seepage quantity was increasing, but that no sediment was observed within the discharge. In addition, decanting of the supernatant pond water from the Surface Cell was now on-going.

On July 15, 1999, mine staff reported that the seepage rate at Thermocouple #13 had slowed down, but that a settlement trough had formed from the crest down to the 376 lift. Numerous longitudinal and transverse cracks were visible. The seepage at Thermocouple #18 had stopped but then started again (at such a slow rate that mine staff was not able to measure it). Water was also observed on the 378 and 380 lifts above Thermocouple #18. No settlement trough was observed, although some cracking was noted.

Since the Surface Cell water level was being lowered, and the quantity of seepage was decreasing, it was determined that seepage event was not exacerbating. Grading of all settled areas was undertaken to prevent any potential erosion resulting from surface run-off. In addition, dike monitoring requirements were outlined, as was a site investigation program to investigate the condition of the dike and the seepage sources.

Two possible mechanisms that may have been responsible for the seepage event were formulated as follows:

Firstly, tailings discharge water ponded against the dike infiltrated into the upstream side of the shale fill, then flowed downward, possibly along the shale fill/tailings contact. The initiating cause in this scenario may have been due to a poor contact of the shale fill to the underlying tailings, possible higher permeability in the unsaturated zone of beached material directly adjacent to the upstream side of the dike and/or cracking of the frozen tailings material leading to seepage. Regardless of the actual mechanism, water seepage was indicated at the bottom of the shale fill as noted earlier.

Secondly, it was alternatively postulated that hydraulic pressures from the impounded water in the deep hole area of the Surface Cell may have been transmitted through the unfrozen zone at depth. Since unfrozen conditions are expected within the tailings proximal to the deep hole, it is possible that hydraulic pressures may have been transmitted through the tailings to the underlying till or fractured bedrock and conveyed downward to the toe of the dike. If extensive frozen conditions at the toe blocked a drainage path for the hydraulic pressures, then artesian pressures may have been created at the toe of the dike.

A review of the seepage event observations, as summarized below, indicated that the first mechanism was responsible for the seepage, rather than the second.

- The deep hole is a considerable distance upstream from the Thermocouple #18 location; seepage water could have only been derived from tailings discharge slurry.
- Once the tailings beach (sealing the upstream side of the dike) advanced south beyond the location of Thermocouple #18, the seepage at this location stopped.
- Once the ponded water level upstream from Thermocouple #13 was lowered so that the pond was not touching the upstream face, then the seepage at Thermocouple #13 stopped. If the second mechanism was possible, then seepage at the toe may have continued, depending upon the hydraulic gradient.
- Tailings sediment was observed within the discharge water, lending credence to the idea that erosion was occurring at the bottom of the shale rockfill layer. The formation of settlement troughs and associated cracking indicates that some internal erosion of the dike did occur.
- Installation of two standpipe piezometers at the toe of dike in August did not indicate any artesian pressures. The second mechanism requires that high hydraulic pressures should be situated at the toe.

Since the first mechanism appeared to be the likely cause of the seepage, a bentonite seal was recommended for the upstream face of the dike. This seal was constructed in the summer of 1999 at the contact of the tailings and upstream dike rockfill, within a 75 to 100 m wide zone around the location of Thermocouple #13. As a result, back-flowing water from the tailings discharge line did not come in contact with the upstream side of the dike when the tailings discharge point advanced past the Thermocouple #13 area. Reports from site indicate that no additional seepage occurred as the discharge point advanced past this area. Therefore, it lends additional credibility to the first seepage mechanism.

Based on thermocouple data obtained on September 2, 1999, after the seepage event of the previous months, significantly more thaw occurred at the section adjacent to Thermocouple #13, as compared to the remainder of the dike.

Given that the West Twin dike was incrementally raised almost every year, it was subjected in summer 1999 to the highest head difference ever (approximately 16 m). Additionally, since the discharge pipe had encroached within 1 m of the crest at the north end of the dike only, this particular lift and location also experienced higher than typical water pressures. Coupled with warmer than typical temperatures within the southern portion of the dike, and possibly some poor bonding (tailings to overlying shale fill) upstream of Thermocouple #18, conditions were created that led to the seepage. Since this event, tighter controls have been placed on the location and elevation of the slurry discharge line.

6 RISK ASSESSMENT

After the events in 1999, it was decided to undertake a risk assessment regarding the stability of the West Twin dike. The primary objective of the assessment was to identify potential failure modes on the dike and to estimate likelihoods of probability, in order to assess the potential risks.

A Failure Mode and Effects Analysis (FMEA) was selected as the appropriate method for this assessment. In this qualitative analysis, the effects or consequences of individual component failure modes are systematically identified. The FMEA is intended to be a formalized method of project review or engineering reliability evaluation which will identify risks and allow the characterization and qualitative ranking of risks. This process does not in itself reduce risks.

A series of failure modes and effects were developed for each of the links and elements identified. A series of potential failure modes were identified based on previous historical data published by United States Committee on Large Dams (USCOLD, 1994) and then extended with potential site specific modes. In total, there were 34 different failure modes identified, although several were repetitive and were distinguished from each other by the consequence results only. Of these 34 potential modes, only one mode was classified as High Risk and one failure mode was classified as Moderately High Risk. Investigative and analytical programs are currently in progress to assess the concerns raised for these two potential modes.

7 CONCLUSIONS

The tailings disposal and dike construction methodologies have been designed within the constraints of the physical requirements of the Arctic and the cost-effectiveness required for mining. An existing natural depression was initially used for storage, supplemented by a solids retention design that used both local materials (as opposed to imported synthetic materials) and permafrost. Additionally, the closure of this facility will again rely on both of this aspects, hopefully achieving a cost effective closure plan.

The upstream constructed, frozen shale dike is a unique design that has performed well to-date, except for occasional seepage events. Some lessons on the its performance have been learned from that event. In addition, the risk assessment process was useful in prioritizing the design elements and operational criteria that apply to the dike, which are noted as follows:

- West Twin dike is designed for the permanent retention of tailings solids only.
- The freeboard amount on the dike will be greater than 2 m, except as follows.
- If the freeboard amount needs to be reduced to less than 2 m, then a suitably-designed sealing element will be placed on the upstream side of the top lift.
- In no case will the freeboard be less than one meter.
- Thermocouples will be used to monitor the subsurface geothermal regime of the dike and its foundation.
- At closure, no pond of water will exist in the Surface Cell, upstream of the dike. An appropriate thickness of shale and armour rock will be placed to ensure that the tailings remain frozen.

These criteria will henceforth be used to guide the exact configuration of the dikes, tailings solids deposition and water levels within the facility.

8 ACKNOWLEDGEMENTS

Appreciation is expressed to Nanisivik Mine, a division of CanZinco Ltd. for permission to publish this paper. In addition, drafting and secretarial support for the production of this paper was supplied by BGC Engineering Inc. Sincere thanks to Gerry Ferris of BGC for his diligent review of the paper.

REFERENCES

B.C. Research 1975. Terrestrial Environmental Studies at Strathcona Sound, N.W.T. Progress Report No. 2 prepared for Strathcona Mineral Services, Project 1552, March, 1975.

International Permafrost Association (IPA), 1997. *Circum-Arctic Map of Permafrost and Ground Ice Conditions*. Map CP-45, edited, by J. Brown, produced by the U.S.G.S., scale 1:10,000,000.

McRoberts, E.C. and Morgenstern, N.R. 1974. The Stability of Slopes in Frozen Soil, Mackenzie Valley, N.W.T. *Canadian Geotechnical Journal*, Volume 11, pp. 554-573.

Nanisivik Mines Ltd., 1993. 1992 Geotechnical Results, Nanisivik Mines, by Robert Carreau (Process Technology Supervisor), 1993.

Nanisivik Mines Ltd., 1994. 1993 Geotechnical Results by John Goyman (Mineral Processing Superintendent), July 17, 1994.

Terratech, 1989. Geotechnical Investigation, Proposed Tailings Containment Scheme, Nanisivik Mine. Report No. 1359-1, submitted to Nanisivik Mines Ltd.

Terratech, 1990. Geotechnical Investigation, Permafrost Aggradation, Nanisivik Mine, NWT. Report No. 1434-1, submitted to Nanisivik Mines Ltd., November 14, 1990.

Terratech, 1993a. Construction Surveillance West Twin Lake Containment Dyke. Report No. 1434-3, submitted to Nanisivik Mines Ltd., January, 1993.

Terratech, 1993b. Geotechnical Test Program, West Twin Lake Test Cells. Report No. 1434-4, submitted to Nanisivik Mines Ltd., January, 1993.

Tordon, Frank, 1997. Hydraulic Confinement of West Twin Lake. Letter to Nanisivik Mine, October 26, 1997.

United States Committee on Large Dams (USCOLD). 1994. Tailings Dam Incidents. November, 1994.

Watts, Griffis and McOuat Ltd., 1973. Feasibility Study of the Strathcona Sound Project. Two volume report submitted to Mineral Resources International Limited.

Mining in the Arctic, Olsen, Lorentzen & Rendal (eds), © 2001 Swets & Zeitlinger, Lisse, ISBN 90 5809 177 5

Foundation of the tower head frame on the "Internationalny" diamond pit

V. U. Izaxon
Yakutsk, North Mining Institute SB RAS, Russia

ABSTRACT: A tower head-frame (Fig.1) at the top of the shaft at the «Internationalny» (Mirny, Yakutia) underground mine has been constructed for transport of men and materials. The foundation of the tower head frame structure is, however, situated in permafrost ground. Frozen ground has no or little bearing strength when thawed, and under thawed condition the ground cannot support the load of the head frame structure. In order to achieve structural stability, often structures are built on piles. The head-frame base under discussion is built on 132 piles, each pile has the load carrying capability of 100 tons. On the basis of mathematical simulation, a freezing system consisting of two concentric freezing holes (Fig.2) have been constructed for re-building and sustaining of frozen state of the foundation base. An automated thermo-measurement system allows controlling of the temperature regime of the tower head-frame foundation.

1 INTRODUCTION

The underground diamond mine «Internationalny» in Yakutia uses a tower head-frame on the cage shaft (Fig.1). This construction is first of its kind in the permafrost regions. The construction applies significant load on the supporting piles. 132 Piles carry the load. The calculated load on each pile is around 100t. Since frozen ground in thawed state has no load bearing capacity, refreezing and sustaining the frozen state of the foundation base during the service life is essential. The frozen state can be maintained by the application of freezing holes. The situation is extremely difficult due to the fact that it is impossible to freeze the entire foundation base completely. It is necessary however to eliminate thawing zone of 0,6-1 m around the support. Thawing of the ground around the support can be eliminated if the phase transformation of the ground can be avoided.

Fig. 1 General view of mine «Internationalny» surface buildings. Tower head-frame is in the center of picture.

Fig.2 Freezing arrangements of the first concentric hole.

2.Mathematical simulation

2.1.Mathematical model of the dynamic temperature field by arbitrary use of seasonal-cooling devices (SCD).

In order to choose optimal parameters for designing the freezing arrangements and for keeping the rock mass around the vertical shaft collar in frozen condition, it is necessary to calculate the temperature regime of strata. There are many variables that have significant influence on the efficiency of the freezing system. The temperature of freezing liquid, diameter of freezing holes, their number and location, ambient air temperature, presence or absence of thermoinsulation of the hole casing and salinity. A solution of the mathematical problem of the heat exchanging process is obtained by numeral method.

In the following a mathematical model of shaft heat exchange between rock mass surrounding the shat and the point freezing holes for an arbitrarily arranged domain in 2-dimension is presented.

The model assumes the solution of 2-dimensional Stephan problem with n heat (sink) discharges points. The equation of the system in polar coordinates is given by:

$$
\begin{aligned}
&\left(c\rho(T)+D\delta(T-T_\phi)\right)\frac{\partial T}{\partial t}=\frac{1}{r}\frac{\partial}{\partial r}\left(\lambda(T)\frac{\partial T}{\partial r}\right)+\frac{1}{r^2}\frac{\partial}{\partial \varphi}\left(\lambda(T)\frac{\partial T}{\partial \varphi}\right)+\\
&+\sum_{i=1}^{i=n}\frac{Q}{r}\delta(r-r_i)\delta(\varphi-\varphi_i);\quad (r,\varphi)\in\Omega, t>0,\\
&-\lambda\frac{\partial T}{\partial r}=\alpha^*(T_\theta-T),\quad r=R_0,\\
&\lambda\frac{\partial T}{\partial r}=,\quad r=b, \qquad\qquad (1)\\
&\lambda\frac{\partial T}{\partial \varphi}\bigg|_{\varphi=0}=\lambda\frac{\partial T}{\partial \varphi}\bigg|_{\varphi=2\pi},\\
&T=T_0(r,\varphi), r_0<r<b, 0<\varphi<2\pi.
\end{aligned}
$$

In the above equation, the variables $c\rho$ and λ are the heat capacity and the thermal conductivity of the rock mass respectively. The variable *D=Lw, where, L is Latent heat of* water, w is density, and α^* is adjustment coefficient of the thermal exchange on the surface which takes into account the presence of thermal isolation. The variable T is air temperature, and Q is the efficiency of the thermal cooling of seasonal-cooling units (SCD), n is the number of SCD, r_i, φ_i *are* coordinates of i–th SCD. The computing zone Ω represents the circle with a radius equal to the radius of thermal influence of shaft b.

The initial conditions for the problem and the boundary conditions are conventional and straightforward.

The air temperature T is changed by harmonic law (Sleptsov V.I. and other, 1996).

$$T = a_l + b_l\sin(c_lt+d_l), \qquad (2)$$

Where, coefficients a_l, b_l, c_l, d_l are obtained by measurement of the climatic conditions in the field.

Heat flow from the cooling devices is determined from temperature of heat-transport medium:

$$Q=\alpha_s\ (T_s-T), \qquad (3)$$

Where, α_s - coefficient of heat exchange on the interface of the SCD-rock mass, T is the temperature of heat transport medium.

Thermal isolation on the surface is calculated as thermal resistance. In that situation, the coefficient of heat exchange in (1) is determined as

$$\alpha^* = \frac{1}{\frac{1}{\alpha_r}+\frac{\delta}{\lambda_r}}, \qquad (4)$$

Where, α_r is the coefficient of heat exchange of rockmasss with air, δ is the thickness of thermal insulation, and λ is the heat conduction of thermal insulating material.

Thermal parameters of the rockmasss such as heat capacity and thermal conductivity are expressed as the summation of the individual parameters values of water and ice.

$$
\begin{cases}
c\rho = (1-m)c\rho_c + m\beta_0(c\rho_B\beta_B + c\rho_{.l}\beta_{.l}), \\
\lambda = (1-m)\lambda_c + m\beta_0(\lambda_B\beta_B + m\lambda_{.l}\beta_{.l})
\end{cases} \qquad (5)
$$

Where, m is porosity, $c\rho_c$, $c\rho_B$, $c\rho_{.l}$ are heat capacity of rockmass, water and ice, β_O, β_B, $\beta_{.l}$ are the initial water saturation and relative water and ice saturation of pore; λ_C, λ_B, $\lambda_{.l}$ - thermal conductivity of rockmass, water and ice, respectively.

The variables, heat capacity and thermal conductivity, take into account of the phase composition change and for phase transmissions of pore saturation. If the mineralization of saturating moisture is considered, and it is assumed that phase transmission from water-ice takes place with the shift in negative temperatures (T<*0)*. The dependence of water saturation on the temperature can be assumed piece-linear for simplification:

$$
\beta_B(T) = \begin{cases} 1, & T > T_\phi, \\ 0, & T < T_\phi, \end{cases}
$$

Where, T *is the* temperature of phase inversion of pore moisture.

Water and ice saturation of pores and unsaturated pores behave as a connected system:

$$\beta_B + \beta_{.l} = 1.$$

We solve the system of governing equations (1) by finite difference method, by discretizing the 2-dimensional equation of heat conduction in time and polar coordinates, and by splitting the governing equation into two 1-dimensional problems by method of total approximation (Samarsky A.A., 1997), thus allowing the solution of one dimensional problem to serve as the initial condition for the other. Solution algorithm of the re-formulated problem results in sequencial solution of 1-dimensional problem for all linear sections of the problem domain for the first and the second directions.

Separation of the main problem into two 1-dimensional problems for radius and circumference, obtained from location points of each of the columns results in increased stability of the differential schemes for determining of the efficiency of heat removal.

3.Construction of the Cooling System.

The following conclusions were made as the result of multivariate simulation experiment:

For heat removal, and for increasing the bearing capacity of the ground, active freezing holes are essential. It is not possible to maintain the temperature regime only with passive seasonal cooling devices.

The desired result was achieved by arranging the active cooling pipes in two concentric cicles around the shaft (Fig.2) and around the piles.

Functioning of the cooling (freezing) system must be controlled by constant temperature measurement and timely activation (turn- on and turn-off) of the cooling devices;

-The parameters of the cooling system are as follows: coolant temperature (-20^{0}C); diameter and depth of the cooling pipes are 0,2m and 70 m respectively.

The Cooling system was assembled and built using the simulation results and the recommendation in 1997 (Fig.3).

4.Construction of the thermometrical system.

For controlling the thermal regime, the temperature measurement were obtained with the help of platinum temperature–sensitive resistors, which are located partly in the vertical holes (20 holes, each 14 m deep – transducers through 1 meter; 8 holes, each 70 m deep– transducers through 2.5 m), and partly in measuring holes, drilled from the shaft (200 holes, each 7 m deep– transducer through 1 m) as shown in Fig.1. Transducers are directly connected with a PC, which transforms the measurement results of each transducer resistance in terms of temperature and prepare an input data file for subsequent use in the calculation in the program.

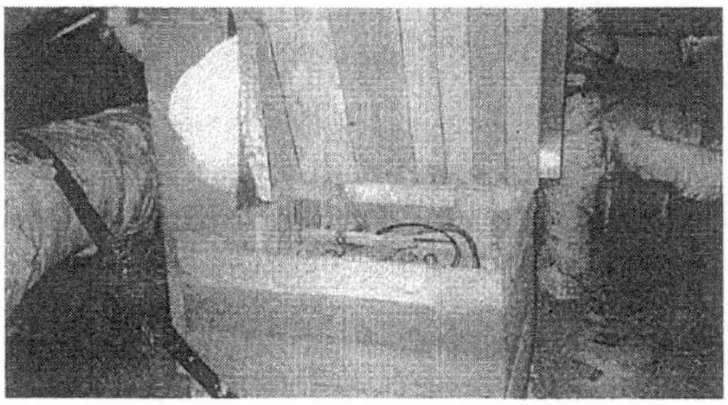

Fig.3 Thermo-insulated head of a measuring hole.

Isotherm maps and load bearing capabilities of the piles, stresses patterns in the shaft support calculated at this stage. Analysis of these results is used to decide changes in the working of the cooling system (continue to work, on or off situation).

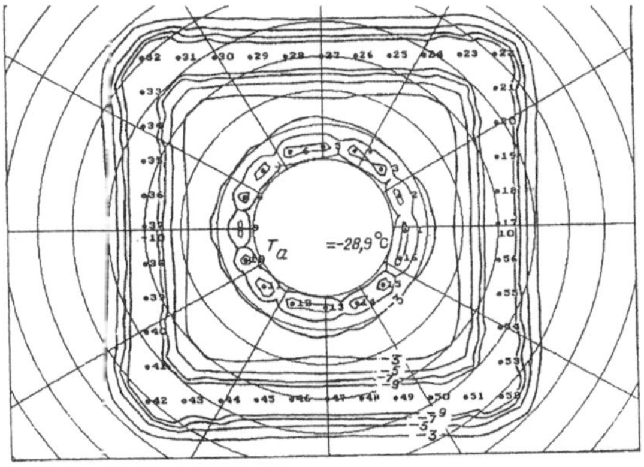

Fig.4. Example of the temperature isotherms

5. Conclusion.

The mathematical model (1) is programmed in Fortran programming language. The program allows to simulation of the temperature field of the base of the tower head frame. Computed results can be seen on the computer screen as well as stored as a data file of with distribution on the screen and in a file of temperature isolines (fig. 4) for subsequent plotting.

The freezing system is in place since 1997. The foundation base of head frame is steady, no significant deformations or failure at the shaft mouth was observed. The temperature measuring system is functioning and is being debugged for more accurate data gathering.

The author acknowledges the support of RFBR for this research.

References

1.Sleptsov V.I., Mordovskoy S.D., Izaxon V.U. Mathematical simulation of heat exchange in permafrost.-Novosibirsk: "Nauka" Publ.House, 1996.- 100 p.

2.Samarsky A.A. Theory of difference schemes.-Moscow: "Nauka" Publ. House, 1997.-656 p.

Mining in the Arctic, Olsen, Lorentzen & Rendal (eds), © 2001 Swets & Zeitlinger, Lisse, ISBN 90 5809 177 5

Engineering control of mine thermal regime around a mine airway in permafrost

S. Bandopadhyay & H. Wu & M. G. Nelson
University Of Alaska Fairbanks, School of Mineral Engineering, Fairbanks, AK 99775

V. Izaxon
Mining Institute of the North, Russian Academy of Sciences, Yakutsk, Russia

ABSTRACT: Regulation of mine thermal regime is extremely important to assure the stability of the mine openings. This paper examines engineering control of mine thermal regime by thermosyphons placed on the surface of the mine. The results indicate that with the thermosyphons installed over the airway roof, a cold area is developed at the top of the roof. The cold area, in some degree, resembles a dome covering over the airway roof. The development and maintenance of this cold dome must be carefully considered in frozen ground when planning the support system of an underground mine opening near the surface.

1 INTRODUCTION

The stability of frozen ground is affected by thawing, which is unusually due to the passing of warm ventilation air through the underground workings. A thermal stream is invariably established between the air passing through the mine and the surrounding rock. This leads to a gradual increase in the temperature of the frozen ground and to a decrease in its ice content.

To assure the stability of mine openings, the temperature of the incoming air needs to be maintained below the freezing point. On the other hand, to assure suitable working conditions for the mine workers and to keep the water lines from freezing, ventilation air must be heated. Regulation of mine thermal regime is thus extremely important, because the thermal regime directly or indirectly affects all elements of mining operations in Arctic.

Without any thermal control, there will be a stable change in temperature along the ventilation path. To prevent such changes, the air entering the mine is often refrigerated when outside air temperatures are above freezing (Udd, 1989). The need for thermal control to ensure stability, including refrigeration of intake air if necessary, was also emphasized by Bandopadhyay and Zhang (1990). If cooling of the intake ventilation air is neither feasible nor economical, subcooling of the permafrost ground would greatly increase the bearing capacity of the rockmass. Ground freezing may be applied to any soil and rock materials regardless of structure, grain size, or permeability. Cryo-Anchors, also referred as thermosyphons and thermotubes, are used throughout Alaska and at many locations in Canada to remove heat and depress the subsurface temperature in permafrost regions. During the Arctic winter season, Cryo-Anchor stabilizers can remove large quantities of heat from the surrounding rockmass, and reject that heat to the air. In the summer, when the air and environment are warmer than the rockmass, the device becomes inactive, minimizing heat input to the ground (Bandopadhyay, et al., 1995). In the following section, engineering control of mine thermal regime around a mine airway with thermosyphons placed on the surface of the mine is presented.

2 ANALYSIS OF THERMAL REGIME AROUND A MINE AIRWAYS WITH THERMOSYPHONS

A physical model of heat transfer around a mine airway near the ground surface with thermosyphons installed is illustrated in Figure 1. On the surface above the mine opening, two rows of thermosyphons are installed parallel to the axis of the opening. Since a single row thermosyphons can be treated as a special case of two rows of thermosyphons, the governing equations were derived based on two rows of thermosyphons placed on the surface. The domain considered, its geometry, and boundary conditions are shown in Figure 2.

The mathematical statement of the heat diffusion problem corresponding to the geometrical configuration shown in Figure 2 can be expressed by the "Stephan's equation":

$$[\rho C_p + L\delta(T - T_0)]\frac{\partial T}{\partial t} = \frac{\partial}{\partial x}[k_x\frac{\partial T}{\partial x}] + \frac{\partial}{\partial y}[k_y\frac{\partial T}{\partial y}]$$

$$+ [Q_1(x,y) + Q_2(x,y)]h_3\ \eta(T_f - T) \ ; \quad (1)$$

$$Q_1(x,y) =$$

$$\delta(y - d_2)\delta(d_2 + a_r - y)\theta(x - \frac{b_3}{2})\theta(\frac{b_3}{2} + a_r - x) \ ; \quad (2)$$

and

$$Q_1(x,y) =$$

$$\theta(y - d_2)\theta(d_2 + a_r - y)\delta(x - \frac{b_3}{2})\delta(\frac{b_3}{2} + a_r - x) \ ; \quad (3)$$

where

a_r : size of the thermosyphon,
d_2 : y-coordinate of the thermosyphon,
T_f : temperature of the working fluid in the thermosyphon,
b_3 : distance between thermosyphons in a row,
δ: Dirac delta function,
h_3: equivalent heat exchange coefficient between the thermosyphon and the frozen ground, and
θ and η : Heaviside functions, which are defined as:

$$\eta(y) = \begin{cases} y & y<0 \\ 0 & y\geq 0 \end{cases},$$

and

$$\theta(y) = \begin{cases} y & y>0 \\ 0 & y\leq 0 \end{cases}.$$

Equation (1) expresses the amount of heat accumulation (right hand side of the equation) in a unit volume equals the amount of changes in heat flow in that volume (left hand side of the equation) with a correction term for a heat sink resulting from the presence of thermosyphons.

2.1 *Initial and boundary conditions*

The initial and boundary conditions for the heat transfer problem around an airway near the surface with thermosyphons, shown in Figure 2, are defined mathematically in the following.

The initial condition (the initial temperature) is

$$T(x,y,t)|_{t=0} = T_0 \ ; \quad (4)$$

The boundary conditions are

(1). On the inside wall, 'A-B', the convective boundary is:

$$-k\frac{\partial T}{\partial y}|_{y=d_1+b_1} = h(T_a - T) \ ; \quad (5)$$

(2). On the inside wall, 'B-C', the convective boundary is:

$$-k\frac{\partial T}{\partial x}|_{x=\frac{b_2}{2}} = h(T_a - T) \ ; \quad (6)$$

(3). On the inside wall, 'C-D', the convective boundary is

$$k\frac{\partial T}{\partial y}|_{y=d_1} = h(T_a - T) \ ; \quad (7)$$

(4). On the top surface, 'E-F', the seasonal air temperature convection boundary is

$$T(x,y,t) = 1 - 10\cos(\frac{2\pi}{365}(t - 18)) \ ; \quad (8)$$

(5). On the far right outside surface, 'F-G', the fixed temperature boundary is

$$T(x,y,t)|_{x\to\infty} = T_\infty \ ; \quad (9)$$

(6). On the bottom surface, 'G-H,' the fixed temperature boundary is

$$T(x,y,t)|_{y\to-0} = T_\infty \ ; \quad (10)$$

(7). On the left symmetrical surfaces, 'H-D' and 'A-E', the insulated boundary is

$$-k\frac{\partial T}{\partial x}|_{x=0} = 0 \ ; \quad (11)$$

(8). The temperature of working fluid in the thermosyphon is

$$T_f = 1 - 10\cos(\frac{2\pi}{365}(t - 18)) \ ; \quad (12)$$

(9). The interface boundary between thermosyphons and the frozen ground is given by

$$k\frac{\partial T}{\partial n}\Big|_{x,\,y\,\in\,\Gamma_t} = h(T_f - T)\ , \quad for \quad T_f < T\ ; \qquad (13)$$

and

$$k\frac{\partial T}{\partial n}\Big|_{x,y\in\,\Gamma_t} = 0, \qquad for \quad T_f \geq T\ ; \qquad (14)$$

where Γ_t is the interface between thermosyphons and the frozen ground.

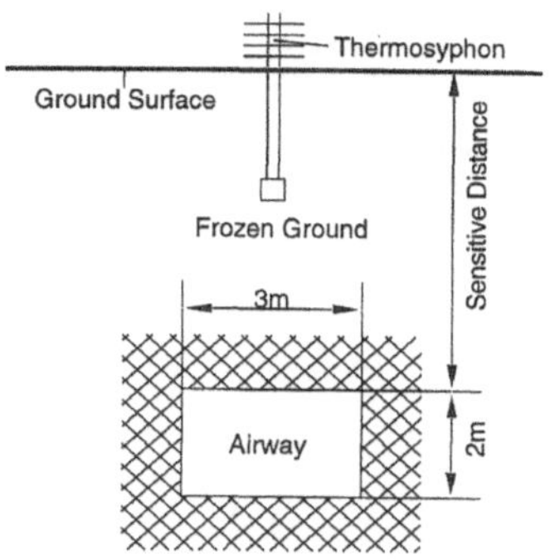

Figure 1: Heat Transfer Around a Shallow Mine Airway With Thermosyphon

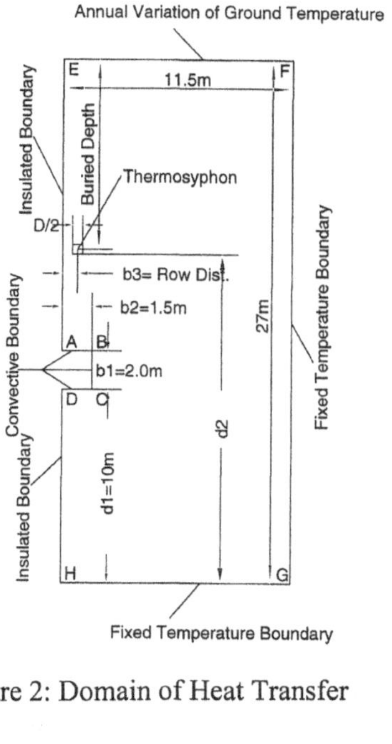

Figure 2: Domain of Heat Transfer

3 SELECTION OF CASES AND PARAMETERS

There are two distinct ways in which the thermosyphons can be installed in relation to the ground surface. A vertical thermosyphon, where the evaporator is imbedded in the frozen rockmass, is installed at nearly a right angle to the ground surface. Its main direction of heat transfer is parallel to ground surface. The horizontal thermosyphon, where the evaporator is placed horizontally or slightly declined, has its main direction of heat transfer along the plane that is perpendicular to the ground surface. For a level or gently declined airway, it is obvious that a horizontal or declined thermosyphon would be more effective. Therefore, only heat diffusion with horizontal thermosyphons was studied here.

3.1 *Finite element solution*

Solutions of the heat transfer problems (equations 1 through 14) were obtained using the ABAQUS (Hibbitt, Karlsson, and Sorensen, Inc.,1992) finite element program. In the ABAQUS analysis, the thermosyphon is simplified as a convective element. Thus, the heat exchange coefficient was taken as the value of the thermosyphon conductance. This is calculated by the following equation:

$$q = h_t(T - T_f) = \frac{C^*\Delta T}{L\pi D}\ , \qquad (15)$$

where

q : rate of heat transfer,

C^*: heat conductance of the thermosyphon,

ΔT: temperature difference between the air and the frozen ground;

L : length of the thermosyphon, and

D : outer diameter of the thermosyphon.

Based on the last two terms of equation (15),

$$h_3 = \frac{C^*}{L\pi D}\ . \qquad (16)$$

With wind velocity of 3 meters/second, C^*=78 W/°C, D=0.18 meters and L=6.1 meters, the value of h_f is calculated as 81.4 W/(m^2-°C).

3.2 *ABAQUS input file and postprocessing*

The finite element mesh for one of the cases studied is shown in Figure 3. To simplify the actual convective boundaries of the thermosyphons, a square element was used to present a thermosyphon and its dimension was taken as 0.3 × 0.3 meters.

The outside diameter of the selected thermosyphon is 0.18 meters.

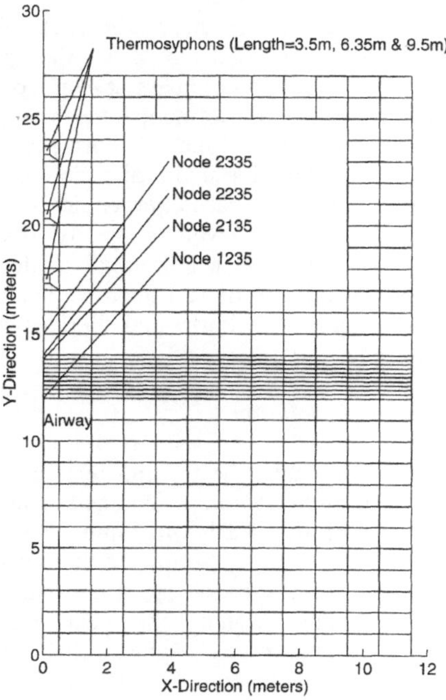

Figure 3: Finite Element Mesh of a Mine Airway with One Row of Thermosyphons

The heat transfer between the thermosyphons and the frozen rockmass was treated as a sink, in which a convective heat transfer was assumed to take place at their interface boundaries. The variation of open air temperature was used as the temperature function of working fluid inside the thermosyphons.

3.3 *Response to annual temperature cycle*

Heat conduction in the soil close to a thermosyphon is primarily in the radial direction. Therefore, for a thermosyphon, vertical heat transfer is relatively unimportant. Soil cooling through the thermosyphon occurs throughout the winter and into the spring. As long as the condenser surface temperature is lower than the soil temperature, heat is extracted directly to the air. Thermosyphons are only effective if they are installed in a sufficient density. Depending on the geometry of mine opening, material characteristics and seasonal air temperature, installation of a series of thermosyphons in rows seems reasonable. In such an installation, distance between rows, the installation depth of the thermosyphon, the geometry of the installation and the spacing of the thermosyphon in the rows are important considerations. In this research, it was assumed that the heat is transferred to the ground from one major source, the ventilating air and the temperature of thermosyphon working fluid is the same as the temperature of open air (equation 14).

To analyze the effectiveness of a thermosyphon in regulating the mine thermal regime, values for several parameters are required. Experimental or reliable data for these parameters are not easily available. The parameter values used in the analysis are shown in Table 1.

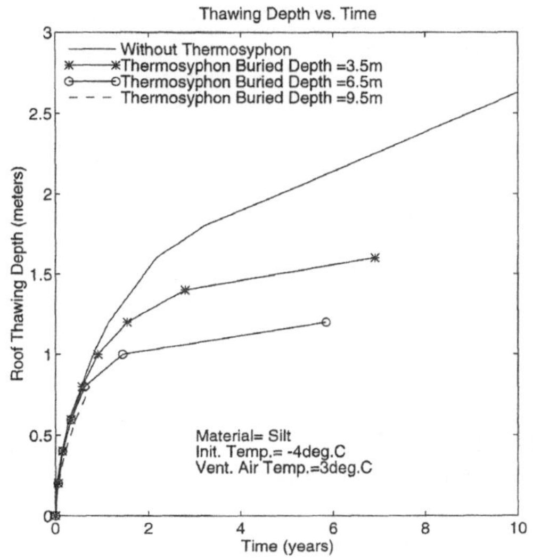

Figure 4: Comparison of Roof Thawing Depths

4 RESULTS AND ANALYSIS

The influence of one row of thermosyphons or sub-cooling of the soil is shown in Figure 4. The depth of roof thawing for thermosyphons burial depths of 3.5, 6.5, and 9.5 meters was reduced by 0.4, 0.8, and 1.2 meters respectively, in comparison with airways without any thermal control. With one row of thermosyphons placed at a depth of 9.5 meters, the depth of thawing approximately 0.8 meters, considerably less than that of an uninsulated roof without thermosyphons.

With two rows of thermosyphons installed over the roof of a shallow airway, the roof thawing was reduced more significantly. With a two meter row separation, the maximum thawing depths observed were 1.2, 0.8, and 0.6 meters, for thermosyphon depths of 3.5, 6.5, and 9.5 meters respectively (Figure 5).

Table 1: Parameter Values used in the Finite Element Calculation

Parameter	Unit	Value
Initial ground temperature	^{o}C	-4
Air temperature	^{o}C	3
Heat transfer coefficient	W/(m^2- ^{o}C)	22.72
Moisture content	%	20
Ice-water latent heat	kJ/kg	30.4
Fine grained soils (silt)		
Thermal conductivity	W/(m-^{o}C)	2.8
Thermal conductivity	W/(m-^{o}C)	1.56
Heat capacity (frozen)	kJ/(kg-^{o}C)	1.13
Heat capacity (thawed)	kJ/(kg-^{o}C)	1.15
Dry density	kg/m^3	1682
Thermosyphon		
Inclination angle	Deg.	12 ^{o}
Thermal conductance	W/^{o}C	78
Diameter of thermosyphon	m	0.18
Number of rows		1 & 2
Length of thermosyphon	m	3.5 6.5 9.5
Dimension of selected	m x m	11.5 x 27
Dimension of mine airway	m x m	3 x 2
Dimension of finite	m x m	1 x 1
Dimension of finite	m x m	0.2 x 1

To illustrate the influence of the distance between rows of thermosyphons, a four meter row distance was arbitrarily selected for the initial study. The maximum thawing depths was 1.2, 0.8, and 0.6 meters with the installation depths of 3.5, 6.5, and 9.5 meters. The changes in the thaw depth caused by this row separation are not meaningful when compared with the resulting thaw with a row separation of two meters (Figures 5 and 6).

This means that with two rows of thermosyphons, some adjustments in row distances can be made without a significant loss in the degree of cooling. The effect of changing the row distances can be seen more clearly in Figures 7, 8 and 9. The roof thawing depths vary little, whether the row separation is tow or four meters. With an increase in the installation depth of the thermosyphons, roof thawing decreased and the time to maximum thawing depth was shorter. The thawing depths for all nine cases analyzed are given in Table 2.

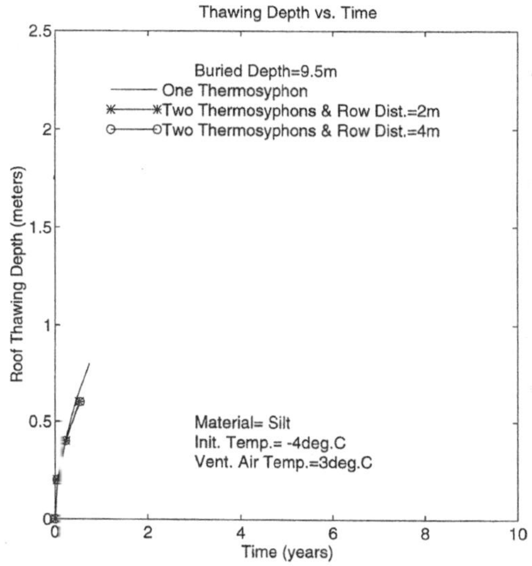

Figure 5: Comparison of Roof Thawing Depths

Table 2: Depths, Time and Amplitude of Thawing Front (with Thermosyphons)

Thermosyphon			Roof Thawing		Amplitude
Case Number	Distance of rows	Buried Depth	Depth	Time	Thawing Front Fluctuations
	m	m	m	years	°C
1	0	3.5	1.8	6	0.11
2	0	6.5	1.2	5.8	0.25
3	0	9.5	0.8	0.7	0.57
4	2	3.5	1.2	1.8	0.15
5	2	6.5	0.8	0.65	0.45
6	2	9.5	0.6	0.3	1.8
7	4	3.5	1.2	2.1	0.2
8	4	6.5	0.8	0.7	0.38
9	4	9.5	0.6	0.5	1.05

Table 3: Sub-cooling Area Over an Airway Roof

Thermosyphon			Summer (9.5 years)		Winter (10 years)	
Case Number	Distance of row	Buried depth	Temperature	Thickness	Temperature	Thickness
	m	m	°C	m	°C	m
1	0	3.5	-2	3.3	-3	2.1
2	0	6.5	-4	5	-7	7
3	0	9.5	-4	5.5	-7	7
4	2	3.5	-5	6.5	-4	10
5	2	6.5	-4	8	-4	12
6	2	9.5	-4	9	-4	13
7	4	3.5	-4	7	-4	10
8	4	6.5	-4	8	-4	12
9	4	9.5	-4	9	-4	13

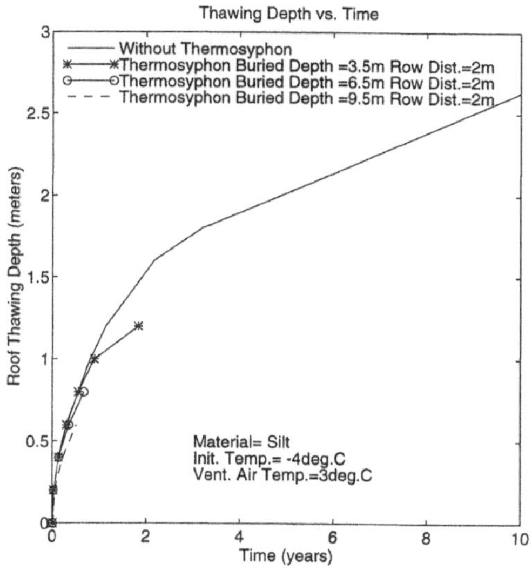

Figure 6: Comparison of Thawing Depth (Row Distance = 4.0 m).

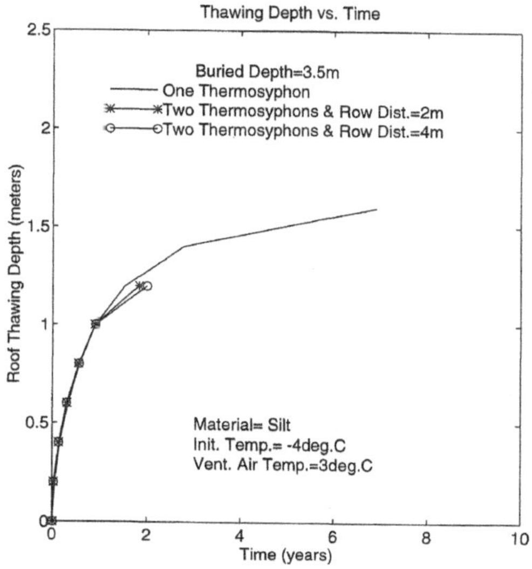

Figure 8: Comparison of Roof Thawing Depth (Length of Thermosyphon = 6.5 m).

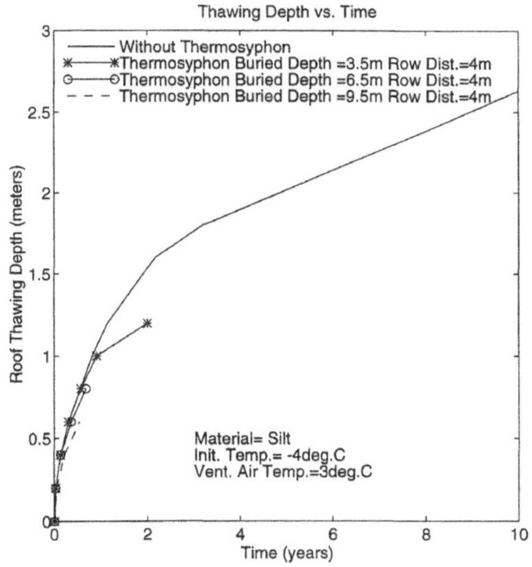

Figure 7: Comparison of Roof Thawing Depth (Length of Thermosyphon = 3.5 m).

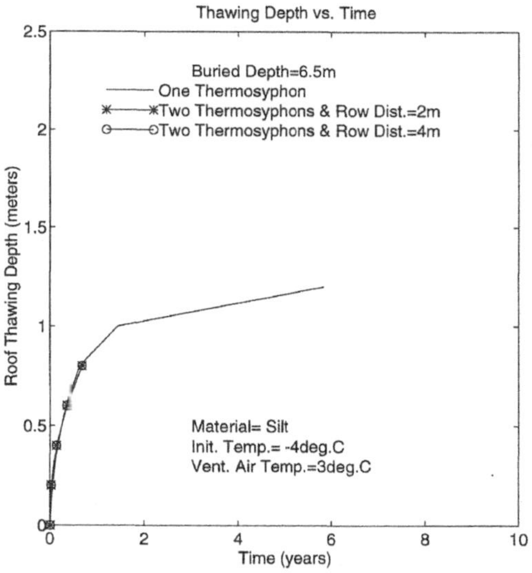

Figure 9: Comparison of Roof Thawing Depth (Length of Thermosyphon = 9.5m).

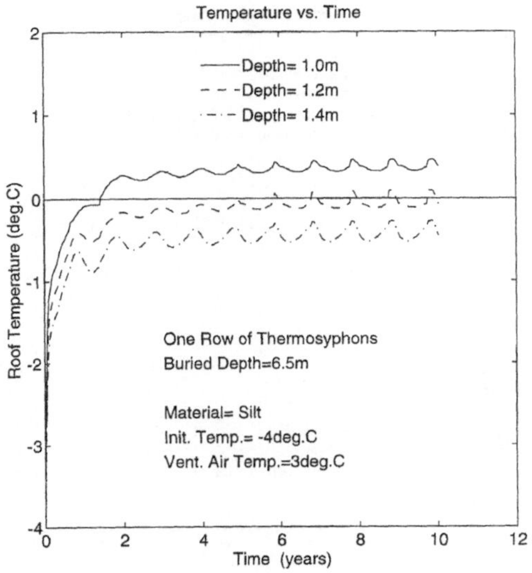

Figure 10: Temperature Variations at Mid-Points on the Roof (Length of Thermosyphon = 6.5 m).

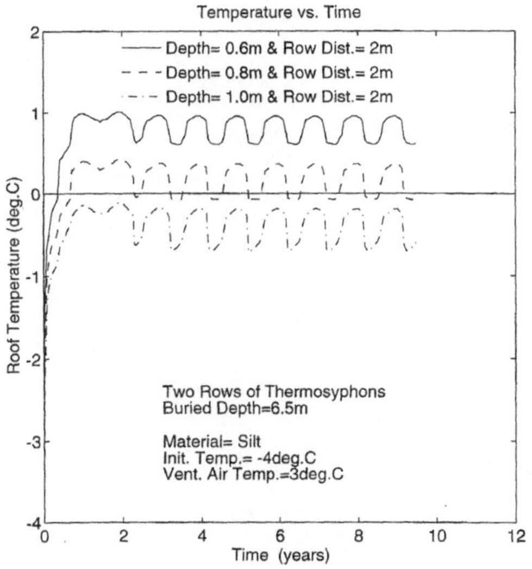

Figure 11: Temperature Variations at Mid-Point on Roof (Length of Thermosyphon = 6.5 m, Row Distance = 2.0).

Figures 10 and 11 are examples of the three selected roof points for each case of one row thermosyphons and two rows of thermosyphons with a row separation of two meters. They were studied and used to show how roof temperature changed in each case. The thawing front node is a node on the top of the airway ceiling, at which the maximum thawing depth reached. The three points are:

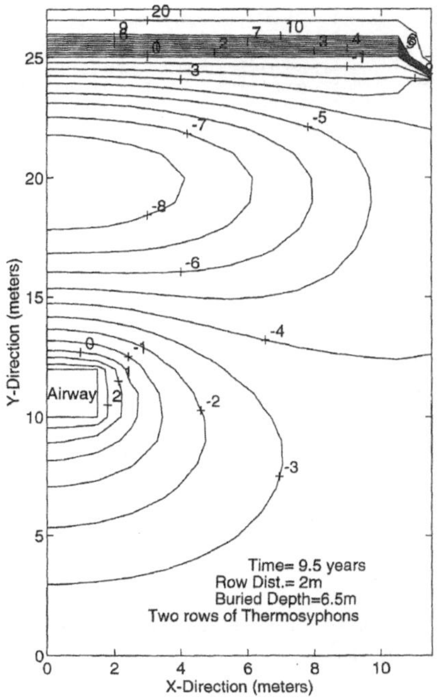

Figure 12: Temperature Distribution Around the Mine Airway.

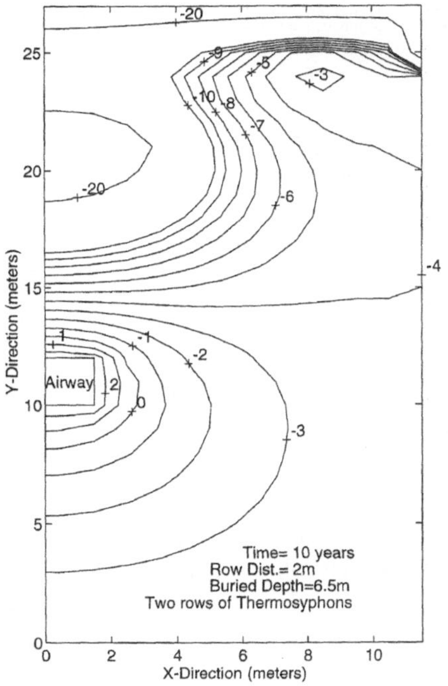

Figure 13: Temperature Distribution Around the Mine Airway.

1. The node just above the thawing front node,
2. The thawing front node, and
3. The node just under the thawing front node.

The fluctuations in seasonal air temperature are reflected by the node temperature variation over the airway roof. The closer the thermosyphon is located to the roof of a shallow airway, the greater the fluctuation of roof node temperature. These fluctuations were due to both the seasonal air temperature variations and the ventilating air temperature. Because of the influence of the air temperature, the thawing front temperature fluctuates slightly around the freezing point even when the temperature distribution approaches a steady status after nearly six years.

With increasing the burial depth, the temperature fluctuation over time increases. For one row of thermosyphons, the amplitude of temperature fluctuation varied from about 0.1 °C, to 0.3 °C and 0.8 °C for burial depths of 3.5, 6.5, and 9.5 meters. With two rows of thermosyphons, both the thawing front retracted and its magnitude of temperature fluctuation over time increased proportionately. When the row distance was two meters, the fluctuations of the thawing front node temperature were 0.25 °C, 0.5 °C and 1.8 °C at respective buried depths of 3.5, 6.5 and 9.5 meters. When the row distance was four meters, the fluctuations decreased to 0.2 °C, 0.35 °C and 1.5 °C at the same respective buried depths. The amplitude of the thawing front node temperature variation is also listed in Table 3. The temperature contours in Figures 12 and 13 illustrate the temperature distributions of two rows of thermosyphons with a separation of two meters at the ends of 9.5 years (summer) and 10 years (winter). With the thermosyphons installed over the airway roof, a cold area developed at the top of the roof. This area was enlarged when the thermosyphon was buried near the airway. And it also became colder. The cold area, in some degree, resembles a dome covering over the airway roof.

Because of the existence of a time lag between the variation of ground surface temperature and the respondence of frozen rockmass, a large, cold area potentially was detained above the airway roof during the summer. Since the frozen soil had larger strength than the thawed or nearly thawing soil, the roof stability was increased.

The thermosyphons caused the upper, cold area to build up during the winter months. The size and depth of this cold area are functions of the open air temperature because of the way by which a thermosyphon works. The thickness of the area varies from about three to 10 meters (Table 3), with the installation depth of 3.5, 6.5, and 9.5 meters and with the one or two rows. The development and maintenance of this cold dome must be carefully considered in frozen ground when planning the support system of an underground mine opening near the surface.

References

Bandopadhyay, S. and Y. Zhang, "Influence of Ventilating Air Fluctuations on Stress Distribution Around an Airway in Frozen Ground," transacytions, SME/AIME, Vol. 288, pp. 1819-1825,1990.

Bandopadhyay, S., H. Wu, M. G. Nelson, and V. Izaxon, "Design of Ventilation Systems for Underground Placer Mines in the Arctic: Analysis of Mine Thermal Regime," Proceedings of the 7th U.S. Mine Ventilation Symposium, A. Wala (Editor), SME/AIME, pp.27-32, 1995.

Hibbit, Karlson and Sorensen, Inc., ABAQUS Users Manuals, Version 5.3, 1992.

Udd, J., "Mining in Canada's North – A Technical Overview, "Proceedings of the 1st International Symposium on Mining in the Arctic, Bandopadhyay, S. and Skudrzyk, F. (Editors), Balkema Publishers, pp. 15-25, 1989.

Mining in the Arctic, Olsen, Lorentzen & Rendal (eds), © 2001 Swets & Zeitlinger, Lisse, ISBN 90 5809 177 5

Ground control and mining in permafrost (including a bibliography)

J. E. Udd & M. C. Bétournay
Mining Laboratories, Natural Resources Canada, Ottawa, Canada

ABSTRACT: Any mine opening which is cut into a rock mass causes redistributions of the pre-existing stress field, alterations to the engineering properties of the geological materials, and the possibilities of failures on either natural or induced fracture surfaces. In the regions in which mining is in permafrost the situation is complicated further through the changes in stability which can result from thawing and re-freezing.

In this paper, the authors review the state of our knowledge, thus far, on this subject. Data from selected case histories are included. An extensive bibliography on references which relate to the ground control implications of mining in the Arctic is appended to the paper.

1. INTRODUCTION

Mankind, in its quest for minerals, has been driven to mine at the limits of its abilities and endurance: from mines near urban centres, with all amenities, to operations in the remotest parts of the globe; from near-surface deposits, which are easily exploited, to those at great depths, at which both rock pressures and temperatures are enormous; from mines at high altitudes in the cordilleras, at which breathing is difficult, to mines at great depths, at which ambient pressures are elevated, and; from mining operations in the hot and humid tropics to those in the frozen regions, at which the opposite extremes of temperatures are experienced.

The limits are defined by accessibility, economics, mankind's ingenuity, and endurance.

In any mining operation, the objectives are to exploit the deposit safely and with an adequate rate of return to the owners.

From a ground control perspective, it must be acknowledged that the act of mining causes disturbances to the conditions which existed at the site previously. The act of excavating a rock mass can result in either disturbances to the pre-existing state of stress or in causing conditions through which failures can take place along pre-existing or induced weaknesses. The objectives of ground control are: (1) to ensure that all of the geological features at the site and the engineering properties of the geological materials have been identified ; (2) to assess how the stability of the rock mass will be affected by the creation of mine slopes and opening ; (3) to provide for monitoring in those parts in which adverse affects can be contemplated ; (4) to suggest either alternative approaches or remedial measures in instances when failure is considered to be likely.

While all of these ground control considerations apply to any mining operation, an additional difficulty of mining in frozen ground is that it is likely that the very act of mining will cause the properties of the geological materials to change. The principal concern is that previously frozen rocks, on thawing, may become appreciably weaker.

In this paper the authors summarize the experiences that have been documented thus far on the ground control considerations which apply to mining in frozen regions.

2. THE DESIGN OF MINING EXCAVATIONS AND SLOPES

One of the most critical steps in the design of any underground mine is the choice of the mining method(s). In this choice a variety of factors come into play, including: the size, shape, attitude, and grade of the mineral deposit; the provision of support that will be necessary; and the costs for personnel, equipment, supplies, energy, and transportation that will be required in order to sustain the operation. In the past, it was not uncommon to find examples of labour-intensive, highly-selective, methods being used for the exploitation of rich deposits. Today, however, because of increasing costs of production, such methods have been largely replaced by lower-cost, bulk methods of mining. Throughout the world there has been a drive towards the implementation of advanced mechanization, automation, and tele-operation, in underground mining methods. The economic "drivers" behind this have been drastically-reduced prices for mineral commodities together with the availability and costs of highly-skilled labour.

In mine design, therefore, there has been a trend to the larger openings which are necessitated by increased productivity and by the sizes of equipment which are required in order to achieve this.

In nature, the size and shape of the orebody imposes many limitation on the choice of a mining method. Pillarless methods, in which a retreat may be made from either the centre or the extremities of the orebody, are generally applicable only to relatively thin tabular deposits (most often of sedimentary origin). In a retreating method the intention is that the back (or roof) will collapse after the mining front has advanced away from it. Local support, such as advancing hydraulic props, are used to ensure that the ground in the immediate vicinity of the mining face is stable. Thus, the concept of the "*critical span*" of the mine opening is of much less concern than in the case of bulk mining methods. In Arctic mining, application of such methods are found in the coal mines on Spitsbergen, in the Norwegian High Arctic.

Bulk mining methods are generally applied to the extraction of thick or massive mineral deposits. These are usually of irregular or lenticular shape, and are often associated with injection on steeply-dipping geological discontinuities. In the exploitation of such deposits some type of stope-and-pillar mining is normally used: with the stopes being extracted in an ordered sequence during the "first pass" of mining; and the intervening pillars being extracted in the "second pass" of mining. Later, remnants, and sill pillars between the mine's levels, are mined in a "third pass", while the mining of the shaft pillar is usually left to the last stages of the life of a mine.

In practice, underground mines are designed with a view towards providing safe operating conditions together with the lowest possible costs of production. The optimum design would feature: the greatest possible separations (intervals) between levels; the largest possible stopes; and the smallest possible pillars in order to maintain stability throughout the period of primary (stope) mining. The "critical span" is the greatest width of opening which can stand unsupported for a given period. It is this which dictates the maximum width of stopes, and whether or not artificial supports, such as cable-bolts, will be required in order to permit even larger dimensions.

The critical span of an opening reflects the strength of the local rock mass and its structural integrity. It is easy to visualize that large unsupported spans will be possible in a massive, strong, unjointed rock mass, and impossible in a weak, badly fractured, material. As regards mining in the frozen regions, then, a fundamental consideration is whether the deposit is in a frozen rock, which loses virtually none of its strength on thawing, or in a frozen material which becomes soil-like on thawing and crumbles easily. At the Black Angel, Lupin, Nanisivik, and Kattiniq mines (Udd and Bétournay, 1999; Pronovost, 1999) the experience has been that the geological materials are strong and quite competent (ie. rock). At Polaris and in the Schefferville open-pit operations of the Iron Ore Company of Canada, however, it has been found that substantial strength is lost when the frozen ores thaw. In such cases the options become either living with the consequences of a deteriorating rock mass or in taking such steps as necessary to maintain stability. These latter may include refrigeration and control of the heat which is produced by the operation of equipment.

Upon the completion of stopes, the customary practice has been to fill the openings with backfill so that support is provided to the rock mass during the mining of the intervening pillars. Historically, as it has become recognized that strong backfill can have an enormous impact on the structural integrity of a mine, there has been a progression from weak natural materials (alluvial sand and waste rock) towards properly "engineered" backfill which provides substantial resistance to deformation. The choice of materials has progressed from: classified mill tailings; to tailings to which cement has been added, in varying quantities (cemented backfill); to high-density paste backfill, to which cement has been added; to cemented rock fill, in which the interstices between pieces of waste rock are filled with cemented backfill, or paste.

A key consideration is that cement is expensive. The costs, particularly transportation, can become prohibitive for remote operations. Thus, at Polaris, frozen waste rock from a quarry on the surface and from underground development was originally used as backfill. With a desire that the backfill should freeze into a solid mass, water was added in order to increase the moisture content. The experience was that from 1.5 to 2 years was required for the heat to be dissipated from the backfill and for it to freeze.

In time, however, it was found that greater strength was required. The result, at Polaris, was a transition to cemented rockfill, with it being necessary to heat the limestone aggregate from the surface quarry in order to prevent the cement slurry from freezing upon coming in contact with the frozen rock.

At Katinniq, however, because of the high costs of transporting cement to the site, the first approach was to use blasthole stope mining together with frozen rock fill, to which water had been added. Unfortunately, because the time to achieve freezing was very long, the choice of a mining method evolved to single-pass cut-and-fill, with each successive cut being completely extracted (Pronovost, 1999).

From the above, it can be appreciated that, from a ground control point of view, the key considerations are whether or not the geological material will deteriorate as the result of the process of mining, and if the management of heat is an important issue as regards both the local deterioration of exposed rock surfaces and the curing of backfilling materials.

Another challenge which may face the ground control engineer is that there may be significant differences in the stability conditions of the different levels of the mine due to the depth at which the local rocks may be frozen and the effects of the geothermal gradient. At many operations it has been noted that the temperatures of the rocks increase with depth. Although these ambient temperatures may still be below freezing it is possible that the increases may result in increased possibilities of instabilities.

Open pit mines cut into frozen ground also present challenges to the ground control engineer. The key consideration is whether or not the slopes will deteriorate significantly as the result of exposure to the elements, and whether failure may result. At the Iron Ore Company of Canada's Schefferville operations (now long closed), the experience was that the exposed rock surfaces thawed and raveled during the months that the ambient temperature was above freezing. On the return of the cold weather, however, the surfaces re-froze.

From a stability point of view, the result of this could be that water is impounded behind the frozen surface layer. With sufficient height of slope, and hydraulic head, the consequence would then be that the surface of the slope is driven to failure. The provision of adequate drainage, therefore, is an important consideration.

The consequences of thawing can be equally serious in small mining operations, such as in placer deposits. In the mid-1990s CANMET assessed the stabilities of the slopes at a number of placers mines in the Yukon Territory (Vongpaisal et al, 1996). From these, it was determined that frozen slopes up to about 20 metres in height, and at angles of up to 60° would probably remain stable during a single mining season.

Drilling and installing support in frozen ground presents special challenges. Since the water normally used will freeze readily the alternatives are to either drill "dry" or to use brine solutions. The former results in excessive dusts, while the latter causes corrosion to both drills and the steel used to fabricate rockbolts, meshes, and cable-bolts.
Care is also required in the selection of either resins or cements which will cure properly and, at the same time, cause minimal damage to either the surrounding rock mass (due to heat) or to the support (due to corrosion).

Finally, blasting practices may also be affected due to the presence of seams and lenses of ice in the rocks. This was particularly noted at the Asbestos Hill mine (now long closed), on Québec's Ungava Pensilula, where extensive breakage of the rocks behind the blasted area and "huge blocks" were often experienced (Udd and Bétournay, 1999).

3. MEASUREMENTS AND MONITORING

During both the design and operating stages of a mining operation it is necessary for a ground control engineer to be aware of the strengths of the local materials, the natural stresses at that location in the earth's crust, and the impacts that mining will have upon both of these. From time to time, it will be necessary to either obtain samples, for testing, or to install instrumentation so that the response of the rock mass to mining can be monitored.

Obtaining and testing samples presents particular challenges. Against the background that severe reductions in strength can result from the thawing of the natural geotechnical materials, an objective should be that the material is tested in its natural state. To accomplish this involves: obtaining samples in their in-situ states; packaging and transporting these in such a manner that these conditions are maintained; and performing the strength tests under controlled ambient conditions (Udd and Pakalnis, 1980; Udd and Yap, 1980). The former will probably necessitate prompt shipment of the samples by air, in well-insulated containers, while the latter will probably necessitate access to a"cold room" or equivalent facilities.

All of these steps are important since the intention should be to obtain the most representative results that one can. An alternative might be to allow the samples to thaw and then to either re-freeze these or to freeze a re-constituted mixture of the material. There are dangers in doing this, however, since the internal structure of the reconstituted and re-frozen material may not approach that of the natural material in-situ, and since the exact properties of the natural ice may be lost.

As regards measurements of the local in-situ stresses in frozen materials, the experience of the Mining Laboratories of Natural Resources Canada is that such tests may be performed successfully - provided that special precautions, appropriate to the ambient conditions, are taken. In particular, a brine solution (22% Calcium Chloride) is required in order to effect the diamond drilling which is required. This, however, results in severe corrosion to both the instruments and the equipment which is used (Arjand and Herget, 1989).

Likewise, the presence of ice in vugs and cavities, in the rock through which the boreholes pass, may result in poor contact between the instrument and the rock. With careful preparation of the rock surface and appropriate adhesives it has been found to be possible to achieve successful bonding between strain cells and the rock surfaces. The Mining Laboratories have used both the USBM (United States Bureau of Mines) and CSIR (Council of Scientific and Industrial Research of South Africa) methods in studies which it performed at the Nanisivik mine (Arjang and Herget, 1989).

It is to be noted that, in Canada, in-situ stress measurements usually indicate that the horizontal tectonic stresses exceed the vertical stresses, which may result from superincumbent loading. At Nanisivik, stress measurements near the east Portal of the mine indicated that the vertical stress was 0.028 MPa/m, while the stress perpendicular to the orebody was 3.2 times as great, and the stress parallel to the orebody was 1.4 times as great.

Concurrently, strain monitoring units, each involving the use of a vibrating wire strain gauge mounted in a large steel ring, were installed in several locations in the mine. The purpose of these monitors is to identify the redistributions of stresses that may result as the pillars are recovered during the final phase of mining. The instruments have been connected to data loggers in order to permit continuous recording of any strain changes.

In addition, ground movement monitors, comprised of linear variable resistors in an aluminum casing, have also been installed in a number of locations in the Nanisivik mine. These latter, which are normally attached to a standard rockbolt, are used widely throughout the mining industry as a means of detecting movements (normally the opening of cracks) along the length of the rockbolt. To the ground control engineer, such movements usually indicate the development of a slab of rock in the back. Any sudden change normally indicates the likelihood of imminent failure (Udd and Bharti, 1987). As of 1999 (Lamond and Auer, 1999), some 29 GMM's had been installed in the mine. Each is connected by cable to a remote terminal unit (RTU) which, in turn, communicates via a leaky feeder modem to a computer on the surface. This is a modern state-of-the-art system. It is to be noted that it has been unaffected by the frozen environment.

4. SUMMARY AND CONCLUSIONS

The objective of ground control is to ensure that the geological materials which surround mining excavations remain stable throughout the mining process and as for long as may be required afterwards The operators of all mines share the same concerns in this respect.

In mines operated in frozen ground, however, additional complications may arise from both the deterioration of the rock mass due to thawing and difficulties in drilling and installing adequate support.

In massive deposits, in which "single pass" mining may not be feasible, some form of stope-and-pillar extraction will be required. In such cases, backfill will be required to ensure that some support is provided during the secondary mining of the pillars. To date, the experience with frozen backfill appears to be that excessively long periods are required in order to the backfill to freeze thoroughly. Thus, increasingly, cemented rock fill is being used.

The following section of this paper contains a bibliography of references which relate to the ground control practices which have evolved at a number of modern mining operations in frozen regions. It is hoped that these will prove to be useful to readers as they search for the finer details of the ground control practices at Arctic mines.

5. BIBLIOGRAPHY OF REFERENCES ON GROUND CONTROL IN ARCTIC MINING

Andres, Dale, "The evolution of pillar mining at the Polaris mine", Proceedings 5th International Symposium on Mining in the Arctic, Edited by John E. Udd and Anthony J. Keen, A. A. Balkema, Rotterdam, Netherlands and Brookfield, Vermont, USA, ISBN 90 5809 098 1, 1999, pp. 125-134.

Anonymous, "Aviation in mineral exploration", in "Report on Mining Operations in the Province of Québec", Department of Highways and Mines, Province of Québec, 1928, Rédempti Paradis, King's Printer, Québec, 1929, pp. 20-21.

Anonymous, " The Black Angel - Canadian mining experience in the High Arctic", Western Miner, October, 1976, pp. 21-24.

Arjang, B. and Herget, G., "Post pillar recovery strategies at Nanisivik mine, Northwest Territories, Phase 1: Ground control instrumentation and stress determinations", Report MRL89-147 (TR), Natural Resources Canada, Ottawa, Canada, November, 1989, 41 pp.

Arjang, B., Herget, G, and Boudreault, A., "Geotechnical site investigations for pillar recovery in permafrost at Nanisivik mine, N.W.Terr.", Proceedings 2nd International Symposium on Mining in the Arctic, Edited by Sukumar Bandopadhyay and Michael G. Nelson, A. A. Balkema, Rotterdam, Netherlands, ISBN 90 5410 078 8, 1992, pp. 75-81.

Arjang, B. and Herget, G, "Post pillar recovery strategies at Nanisivik Mine, Northwest Territories: Phase I ground control instrumentation and stress determinations", CANMET Report MRL89-147 (TR), Energy, Mines and Resources Canada, 1989.

Bandopadhyay, S.; Ke, J.; Nelson, M. G.; Chen, G. and Izaxon, V., "Analysis and prediction of water infiltration in underground, frozen placer mines", Proceedings 5th International Symposium on Mining in the Arctic, Edited by John E. Udd and Anthony J. Keen, A. A. Balkema, Rotterdam, Netherlands and Brookfield, Vermont, USA, ISBN 90 5809 098 1, 1999, pp. 71-78.

Bandopadhyay, S.; Wu, H.; Nelson, M. G. and Izaxon, V., "A finite element model of heat transfer in shallow placer mine airway in the Arctic", Proceedings 5th International Symposium on Mining in the Arctic, Edited by John E. Udd and Anthony J. Keen, A. A. Balkema, Rotterdam, Netherlands and Brookfield, Vermont, USA, ISBN 90 5809 098 1, 1999, pp. 135-145.

Bauer, A.; Harris, G. R.; Lang, L.; Preziosi, P. and Selleck, D. J., "Review of Iron Ore Company's high-explosive cratering program", Proceedings, Rock Mechanics Symposium, University of Toronto, January, 1965, Department of Mines and Technical Surveys (now Natural Resources Canada), Ottawa, Canada, catalogue M37-965, 1965, pp. 107-119.

Bétournay, M.C., "High post pillar recovery in permafrost conditions: survey and discussion of existing information", Report MRL89-24 (TR), Natural Resources Canada, Ottawa, Canada, January, 1989, 25 pp.

Brugmans, Peter J., "Spitsberfen - mining coal with the Arctic Circle", Mining Magazine, London, England, June, 1987, ISSN 0308-6631, pp. 479-487.

Brugmans, P. J., "Major aspects of the operation of coal mines in/below the permafrost area on Svalbard", Proceedings 1st International Symposium on Mining in the Arctic, Edited by Sukumar Bandopadhyay and Frank J. Skudrzyk, A. A. Balkema, Rotterdam, Netherlands, ISBN 906191 899 5, 1989, pp. 87-93.

Bullis, H. R., "Exploration in arctic environments", Proceedings 2nd International Symposium on Mining in the Arctic, Edited by Sukumar Bandopadhyay and Michael G. Nelson, A. A. Balkema, Rotterdam, Netherlands, ISBN 90 5410 078 8, 1992, pp. 35-38.

Cathro, D. C., Hayley, D. W., and Keen, A. J., "Design and construction of a permafrost core earth dam", Proceedings 2nd International Symposium on Mining in the Arctic, Edited by Sukumar Bandopadhyay and Michael G. Nelson, A. A. Balkema, Rotterdam, Netherlands, ISBN 90 5410 078 8, 1992, pp. 153-162.

Diment, T. and Dismuke, S., "The Polaris cemented rockfill project", Proceedings 98th CIM Annual General Meeting, Edmonton, Alberta, Canadian Institute of Mining, Metallurgy and Petroleum, Montréal, Canada, 1996, CD-ROM.

Dismuke, Steve and Diment, Tom, "The testing, design, construction, and implementation of cemented rockfill (CRF) at Polaris", CIM Bulletin, Canadian Institute of Mining, Metallurgy and Petroleum, Montréal, Canada, ISSN 0317-0926, Vol. 89, No. 1005, November/December, 1995, pp. 91-97.

Dorman, K. and Gooch, A. E., "Spray-applied polyurethane foam to insulate heated rooms excavated in permafrost", United States Bureau of Mines, Report of Investigations RI 7392, 1970.

Drake, J. S., and Keohane, P. G., "The Polaris Mine", CIM Bulletin, Canadian Institute of Mining, Metallurgy and Petroleum, Montréal,

Canada, ISSN 0317-0926, Vol. 78, No. 880, August, 1985, pp. 49-53.

Dubnie, Amil, "Northern mining problems with particular reference to unit operations in permafrost", Mines Branch Technical Bulletin TB 148, Energy, Mines and Resources, Canada, Information Canada catalogue number M34-20/148, 1972, 18 pp.

Duval, D., "Innovations in permafrost drilling at Nanisivik", Northern Miner, ISSN 0029-3164, 1979 05 31, p. 12.

Eichanninow, E. A.; Shor, A. I. and Rozenbaum, M. A., "The stability of underground workings in permafrost", Proceedings 3rd International Conference on Permafrost, 1978, Vol. 1, pp. 922-927.

Eigenbrod, K. D. and Roggensack, W. D., "Thermal considerations for mine development and backfill design, Polaris mine", Proceedings 36th Canadian Geotechnical Conference, 1983.

Espley, G. H., "Experience with permafrost in gold mining", Proceedings 3rd Canadian Conference on Permafrost, 1969, pp. 59-64.

Fangel, H., "Use of ice fill is being developed as an important new mining technique", Northern Miner, ISSN 0029-3164, 1975 06 19.

Fish, Richard, "Mining in Arctic lands...the Black Angel experience", Canadian Mining Journal, ISSN 0008-4492, Vol. 95, No. 8, August, 1974, pp. 26-36 (Includes "Looking over the Black Angel's shoulder", by Eric Mikkelborg, pp. 30-31).

Fish, Richard, " Nanisivik: the place where people find things", Canadian Mining Journal, ISSN 0008-4492, Vol. 100, No. 9, September, 1979, pp. 34-48.

Giegerich, H. M., "Arctic mining in permafrost", Proceedings 5th International Conference on Permafrost, Trondheim, Norway, 1988, pp. 1382-1387.

Giegerich, H. M., "Keynote address: Mining in the Arctic", Proceedings 2nd International Symposium on Mining in the Arctic, Edited by Sukumar Bandopadhyay and Michael G. Nelson, A. A. Balkema, Rotterdam, Netherlands, ISBN 90 5410 078 8, 1992, pp. 3-9.

Gorski, B., "Frozen backfill strength determinations, Polaris Mine", CANMET Report MRL91-039 (TR), Energy, Mines and Resources Canada, 1991.

Green, Charles B. and Bundtzen, Thomas K., "An overview of current mineral exploration, development and production in Alaska", Proceedings 1st International Symposium on Mining in the Arctic, Edited by Sukumar Bandopadhyay and Frank J. Skudrzyk, A. A. Balkema, Rotterdam, Netherlands, ISBN, ISBN 906191 899 5, 1989, pp. 205-290.

Hopkins, John L., "Beyond Resolute", Canadian Mining Journal, ISSN 0008-4492, Vol. 107, No. 8, August, 1986, pp. 12-15.

Horne, M. F., "Open pit operations in permafrost, Giant Yellowknife Mines Limited", Paper Number 87,CIM Annual General Meeting, Canadian Institute of Mining, Metallurgy and Petroleum, Montréal, Canada, 1977, 10 pp.

Hull, J. A.; Stuckert, B. J. and Sullivan, J., "Design of paste tailings disposal in the Russian Sub-Arctic", Proceedings 5th International Symposium on Mining in the Arctic, Edited by John E. Udd and Anthony J. Keen, A. A. Balkema, Rotterdam, Netherlands and Brookfield, Vermont, USA, ISBN 90 5809 098 1, 1999, pp. 147-153.

Keen, A. J., "Polaris update", CIM Bulletin, Canadian Institute of Mining, Metallurgy and Petroleum, Montréal, Canada, ISSN 0317-0926, Vol. 85, No. 961, June, 1992, pp. 51-57.

Keen, A. J., "Tailings disposal at Polaris mine", Proceedings 4th International Symposium on Mining in the Arctic, Edited by Arne M. Myrvang and Johannes Vik, SINTEF Rock and Mineral Engineering, Trondheim, Norway, ISBN 82-595-9660-1, 1996, pp. 17-24.

Kilgour, R. J., "Mining experience with permafrost:, Proceedings 3rd Canadian Conference on Permafrost, 1969, pp. 65-70.

Knight, G., Boyle, R. and Boutin, M., "Computer simulation of heat flow in freezing fill in cold mines", CANMET Report MRL89-45 (TR), Energy, Mines and Resources Canada, 1989.

Lamond, Rod and Auer, Luc., "Post pillar mining at Nanisivik mine", Proceedings 101st CIM Annual General Meeting, Calgary, Alberta, 1999, Canadian Institute of Mining, Metallurgy and Petroleum, Montréal, Canada, CD-ROM.

Lang, L. C., "Blasting frozen iron ore at Knob Lake", Canadian Mining Journal, ISSN 0008-4492, Vol. 87, No. 8, August, 1966, pp. 49-53.

Lang, L. C., "New permafrost blasting method developed at Asbestos Hill", Canadian Mining Journal, ISSN 0008-4492, Vol. 97, No. 3, March, 1976, pp. 48-53.

Mian, M. Naseem and Ehrsam, Curtiss O, " 'Northern exposure' the Red Dog experience", CIM Bulletin, Canadian Institute of Mining, Metallurgy and Petroleum, Montréal, Canada, ISSN 0317-0926, Vol. 89, No. 1005, November/December, 1996, pp. 87-90.

Mongeot, C., Gorski, B., Herget, G., and Burn, C.R., "Quasi-frozen high pit wall stability analysis of placer mines, Yukon Territory: Phase I. Indian and Northern Affairs Canada, Northern Affairs: Yukon Region", Open File 1993-6 (T), Volume II-Technical Reports, 1993.

Morken, Per ole and Addison, Thomas B., "Cola pillar extraction in Mine #7, Svalbard. Practical aspects", Proceedings 4th International Symposium on Mining in the Arctic, Edited by Arne M. Myrvang and Johannes Vik, SINTEF Rock and Mineral Engineering, Trondheim, Norway, ISBN 82-595-9660-1, 1996, pp. 239-247.

Moss, A.; Scoble, M. and Matthews, K., "rock mechanics and production management", Proceedings 97th CIM Annual General Meeting, Halifax, Nova Scotia, Canadian Institute of Mining, Metallurgy and Petroleum, Montréal, Canada, 1997.

Myrvang, A. M. and Utsi, J., "Rock mechanics investigations in the Svea coal mine, Svalbard", Proceedings 1st International Symposium on Mining in the Arctic, Edited by Sukumar Bandopadhyay and Frank J. Skudrzyk, A. A. Balkema, Rotterdam, Netherlands, ISBN 906191 899 5, 1989, pp. 95-103.

Nantar Engineering Limited, "The behaviour of frozen backfill in underground mines", Final Contract Report, Energy, Mines and Resources Canada (now Natural Resources Canada), Ottawa, Canada. 1983.

Nantar Engineering Limited, "Time dependent behaviour of frozen pelletized tailings in underground mines", Final Contract Report, CANMET Contract 26SQ.234440-4-9129, Energy, Mines and Resources Canada (now Natural Resources Canada), Ottawa, Canada, 1985.

Price, Ray, "Yellowknife", Peter Martin Associates, Toronto, Ontario, 1967, p. 22.

Pronovost, F., "Arctic mining in Raglan", Proceedings 5th International Symposium on Mining in the Arctic, Edited by John E. Udd and Anthony J. Keen, A. A. Balkema, Rotterdam, Netherlands and Brookfield, Vermont, USA, ISBN 90 5809 098 1, 1999, pp. 15-28.

Pugsley, T. F. and Gignac, D., "Raglan project optimization and economics", CIM Bulletin, Canadian Institute of Mining, Metallurgy and Petroleum, Montréal, Canada, ISSN 0317-0926, Vol. 89, No. 1005, November/December, 1996, pp. 84-86.

Renshaw, Alan E., "Practical blasting in permafrost", Proceedings 4th International Symposium on Mining in the Arctic", Edited by Arne M. Myrvang and Johannes Vik, SINTEF Rock and Mineral Engineering, Trondheim, Norway, ISBN 82-595-9660-1, 1996a, pp. 121-127.

Renshaw, Alan E., "Engineering techniques to optimize an Arctic mine", CIM Bulletin, Canadian Institute of Mining, Metallurgy and Petroleum, Montréal, Canada, ISSN 0317-0926, Vol. 89, No. 1005, November/December, 1996, pp. 77-83.

Seymour, J. B., Tesarik, D. R., and McKibbin, R. W., "Stability of permafrost gravels in an Alaskan underground placer mine", Proceedings 4th International Symposium on Mining in the Arctic, Edited by Arne M. Myrvang and Johannes Vik, SINTEF Rock and Mineral Engineering, Trondheim, Norway, ISBN 82-595-9660-1, 1996, pp. 147-156.

Sims, J. F. M., "An overview of coal mining in Alaska", Proceedings 1st International Symposium on Mining in the Arctic, Edited by Sukumar Bandopadhyay and Frank J. Skudrzyk, A. A. Balkema, Rotterdam, Netherlands, ISBN 906191 899 5, 1989, pp. 3-8.

Stjern, Gisle, Ruistuen, Helge, and Myrvang, Arne, "Performance of rock bolts in the Svea coal mine, Svalbard", Proceedings 4th International Symposium on Mining in the Arctic, Edited by Arne M. Myrvang and Johannes Vik, SINTEF Rock and Mineral Engineering, Trondheim, Norway, ISBN 82-595-9660-1, 1996, pp. 157-168.

Taylor, A. and Judge, A., "Mining companies assist in studies of permafrost in northern Quebec", Northern Miner, ISSN 0029-3164, November, 1978, p. 15.

Udd, John E. and Pakalnis, V., Jr., "The strength of frozen ore in shear", Preprints, 2nd International Symposium on Ground Freezing, The Norwegian Institute of Technology, Trondheim, Norway, 1980, pp. 297-308.

Udd, John E. and Yap, S.-M., "Strength reductions due to the thawing of frozen ores", Preprints, 2nd International Symposium on Ground Freezing, The Norwegian Institute of Technology, Trondheim, Norway, 1980, pp. 309-324.

Udd, J. E. and Bharti, S., "Monitoring of stability conditions at Falconbridge's Strathcona mine", CIM Bulletin, Canadian Institute of Mining, Metallurgy and Petroleum, Montréal, Canada, ISSN 0317-0926, Vol. 80, No. 908, December, 1987, pp. 46-53.

Udd, John E., "Mining in Canada's north - a technical overview", Proceedings 1st International Symposium on Mining in the Arctic, Edited by Sukumar Bandopadhyay and Frank J. Skudrzyk, A. A. Balkema, Rotterdam, Netherlands, ISBN 906191 899 5, 1989, pp 15-25.

Udd, John E. and Bétournay, Marc C., "Some ground control considerations for mining in permafrost", Proceedings 5th International Symposium on Mining in the Arctic, Edited by John E. Udd and Anthony J. Keen, A. A. Balkema, Rotterdam, Netherlands and Brookfield, Vermont, USA, ISBN 90 5809 098 1, 1999, pp. 43-51.

Van der Walt, J., "Cooling the world's coldest mine", Journal, Mine Ventilation Society of South Africa, Vol. 37, No. 12, 1984, pp. 138-142.

Vik, J. S., "Coal mining on Svalbard, Spitsbergen",Proceedings 5th International Symposium on Mining in the Arctic, Edited by John E. Udd and Anthony J. Keen, A. A. Balkema, Rotterdam, Netherlands and Brookfield, Vermont, USA, ISBN 90 5809 098 1, 1999, pp. 155-160.

Vongpaisal, S., Udd, J.E. and Larocque, G.E., "Post pillar recovery strategies at Nanisivik Mine: Phase 2, Ground stability evaluation", Report MRL90-27 (TR), Natural Resources Canada, Ottawa, Canada, April, 1990, 51 pp.

Vongpaisal, S.; Herget, G.; Udd, J. E.; Larocque, G. E. and Bailey, F., "Post pillar recovery strategies at Nanisivik mine, Northwest Territories, Phase 3: Pillar recovery trial", Report MRL91-123 (TR), Natural Resources Canada, Ottawa, Canada, November, 1991, 26 pp.

Vongpaisal, S., Herget, G., Udd, J. E., Larocque, G. E., McNeil, W. H., McConnell, J. C., Boudreault, A., and Bailey, F., "A strategic analysis of post-pillar recovery at Nanisivik mine", Proceedings 2nd International Symposium on Mining in the Arctic, Edited by Sukumar Bandopadhyay and Michael G. Nelson, A. A. Balkema, Rotterdam, Netherlands, ISBN 90 5410 078 8, 1992, pp. 89-95.

Vongpaisal, S., Herget, G., Gorski, B. and Mongeot, C., "Quasi-frozen high pit wall stability analysis of placer mines, Yukon Territory: Phase I. Indian and Northern Affairs Canada, Northern Affairs: Yukon Region, Open File 1993-6 (T), Volume I-Summary, 1993.

Vongpaisal, Somchet, Udd, John, Herget, Gerhard, Gorski, Bernie, Prasad, Naresh, and Mougeot C., "Guidelines for the design of stable slopes for placer mines operating in permafrost", Proceedings 4th International Symposium on Mining in the Arctic, Edited by Arne M. Myrvang and Johannes Vik, SINTEF Rock and Mineral Engineering, Trondheim, Norway, ISBN 82-595-9660-1, 1996, pp. 103-109.

Wilson, Hugh R., "Tailings management program - An operating success, Echo Bay's Lupin mine", Proceedings 1st International Symposium on Mining in the Arctic, Edited by Sukumar Bandopadhyay and Frank J. Skudrzyk, A. A. Balkema, Rotterdam, Netherlands, ISBN 906191 899 5, 1989, pp. 135-144.

Wilton, D. H. C., Archibald, S. M., Hussey, A. M., and Butler, R. W., Jr., "Metallogeny of the Ramah Group: Discovery of a new Pb-Zn exploration target, Northern Labrador", Current Research (1994), Geological Survey Branch, Newfoundland Department of Mines and Energy, Report 94-1, pp. 415-428.

Zahovskis, C. N., "Pillar mining at Polaris", Proceedings World Mining Congress, Stockholm, Sweden, 1987.

Mining in the Arctic, Olsen, Lorentzen & Rendal (eds), © 2001 Swets & Zeitlinger, Lisse, ISBN 90 5809 177 5

Access road across Høganes glacier to the new Svea mine, Svalbard

P. O. Morken
Store Norske, Longyearbyen, Norway

ABSTRACT: The road is devided in 3 sections: 1) The first 2500 m in the beach-zone, done mainly by waste from the mine. 2) The next 1000 m. Enters with a big filling on the top of the end-moraine in front of the glacier. 3) The last 3500 m are on the mid-moraine and on the glacier. The road was completed with filling material, mainly from a split flow river and end-moraine, in the period from May 1999 to March 2000. The filling material consists of 350,000 m^3, and totally 100,000 m^3 of ice are removed from top of mid-moraine. The costs so far are totally 19 mill or 1580 kr/m. The road has a width of approximately 12 m, well built up (2-4m) from the terrain. At the moment it remains to put a 40 cm top layer of gravel. This is planned to be done this summer, when the major settlements in the material and base are calmed down. Even though the movements of the glacier in periods are measured quite high, the insulated road are not moving significant.

MAIN ACTIVITIES IN STORE NORSKE UP TO LONGWALL START UP IN SVEA NORD.

Table 1.

	1999	2000	2001	2002	2003	2004
Remaining operation in mine 7 (2 CM's running in 1999, 1 CM from 2000)						
Remaining R&P operation in Svea mine						
Political clarification on opening Svea Nord						
Upgrading of road and area outside mine (Consisting of: stripping the coal seam clear from ice, upgrading the road, establishing the area outside mine.)						
Complete development works before longwall assembly (Mainly 2 CM's).						
R&P-operation in Svea-Nord (1 CM operating from 01.08.00 - 01.01.01.)						
Assembly of longwall equipment						
Start up of the first longwall panel						

Alternative 3: Optimum rate of development works in Svea Nord.
Production

Table 2. Key Elements/Performance

Key elements	Performanced
Total length of the road	7,000 metres
Filling material and sub-case	212,500 m^3
Big filling from river delta to end-moraine of the glacier	96,000 m^3
Top layer	55,000 m^3
Total amount filled in	363,500 m^3
Total amount removed ice from top of mid-moraine	100,000 m^3
Costs totally (NOK)	19,000,000,-
Additional costs for top layer (summer 2001)	5,000,000,-
Costs/meter	3,429,-
Costs/m^3	66,-
Expected annually maintenance costs for the road excluding removal of snow (kr. 5,-/ton)	6 mill.

Main steps in the 7.000 meter road from Svea to the mine opening

1. The first 2.500 meter of the road, in the beach, is build in the period 01.05.99 to 01.08.99. The filling material is purely waste from the mine.
2. The next 1.000 meter goes through a river split flow area and enters with a big filling on top of the end-moraine in front of the glacier. The filling materials are a mixture of material from the river split flow area and the end-moraine.
3. The last 3.500 meter is mainly on the mid-moraine, except the last 400 meter, which cross over the glacier (towards the mine opening). The filling material are mainly from the end moraine. Step 2 and 3 were built in the period 01.08.99 to 20.12.99. This summer the 40 cm top gravel will be carried out on the top of a geo-textile.

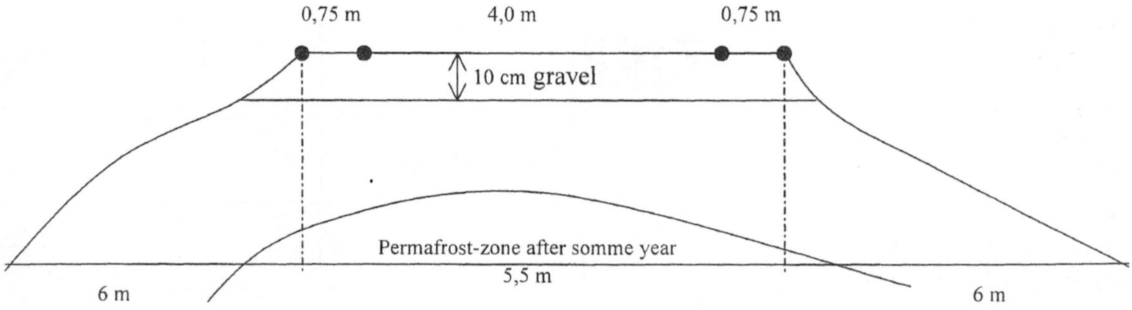

Figure 1. Original planned cross section of the road

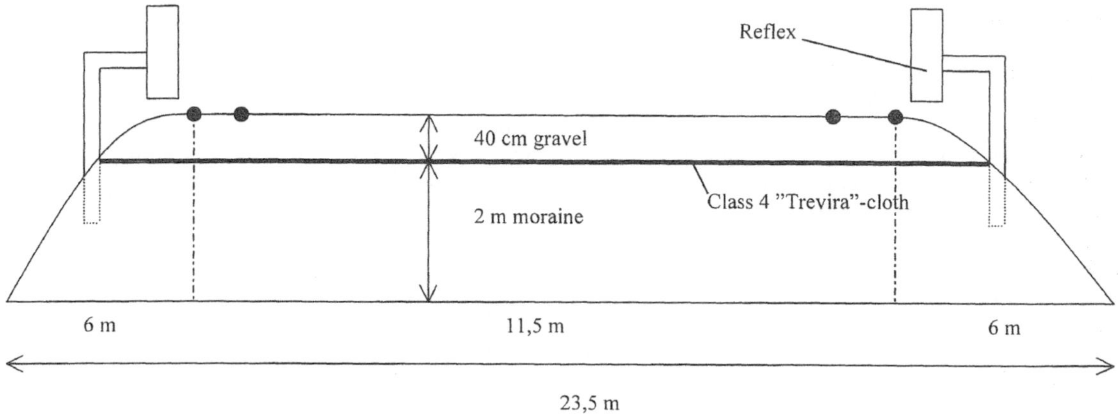

Figure 2. Final cross section of the road

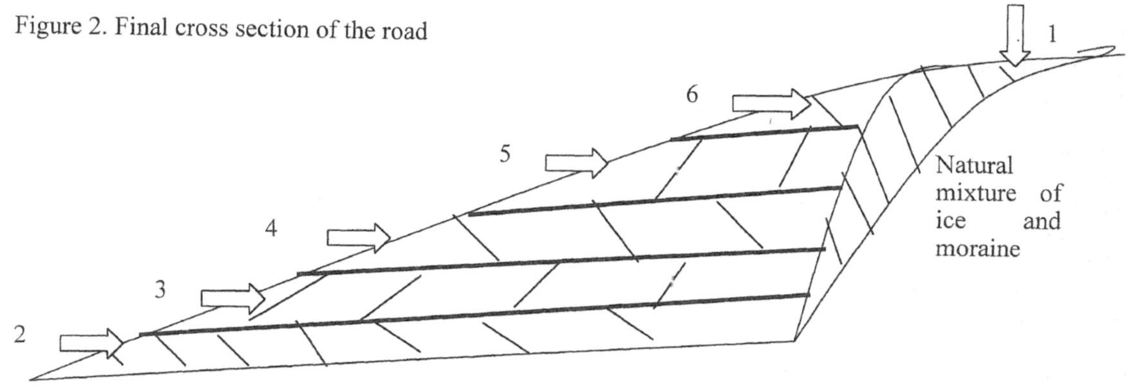

- 96.000 m^3 totally
- 196 m wide in bottom (at widest)
- 23 m height by the highest point

Figure 3. Building the big filling in front of end-moraine

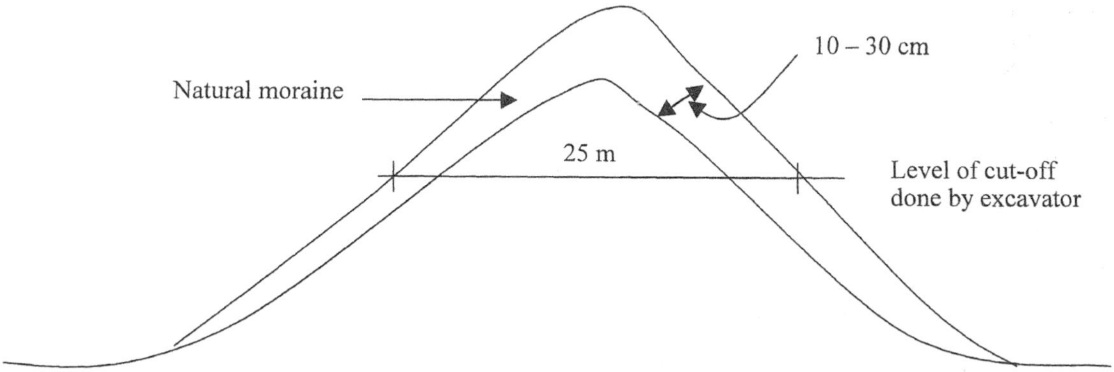

Figure 4 Principal sketch of mid-moraine

Experience and future challenges with the road

Settlements in the beach-zone. It has so far twice been necessary to put additional filling material on the road.

Regional movements on the mid-moraine and glacier. 3 series of measurements of movements of the actual glacier have been performed. This shows annual movements typically like 0,5 meter vertically (melting), and 1,5 meter horizontally. Where the road is, the movements are more like 0,2 meter vertically and 0,8 meter horizontally.

Water erosion in the upper 400 meter of the road (across the glacier). This phenomena has been less than expected. The reason for this is because we drain the water through an excavated channel on the upper side of the road, parallel to the road.

Water erosion in a particularly critical point between end-moraine and mid-moraine (these two points are most critical in the heaviest melting period from beginning of July to end of August). Also this erosion problem has not turned out to be as bad as expected.

Build up a layer of moraine beside the road to insulate in order to prevent big melting of the glacier. This when it comes to the last 400 meter towards the mine opening. The purpose is to avoid "under-digging" by melted ice and water too close to the road.

This summer (2001) we believe that the major settlements are finished. We will therefore put the top level of 40 cm of gravel on the road. Find a suitable mixture to put in the top gravel zone to avoid dry-out and wind-erosion of the fines (measure taken in order not to accelerate the melting of the glacier).

In some parts of the road we have to put in getextile before the top gravel layer is added. This is the part of the road where the filling material consists of too much fines.

Avoid any build ups beside the road for the purpose of minimizing mechanical removal of snow. (This is a special problem with drifting, dry snow.) Some adjustments have been necessary; in some places we have removed peaks in the terrain beside the road, in some other places we have built up the road.

Pay attention to and decide if we have to put up any prevention facilities against avalanche in one particular area of the road. This particular area has been exposed to avalanches at least once a year before the road was built. Since the road was built we have not had any, but we pay attention to it, specially when the weather is mild with heavy snow. We will use helicopter to place explosives if the conditions say we have to create a controlled avalanche.

Use of the experience of the trucks on the road so far when it comes to future demand for big trucks. So far we have been using 40 and 70 tones Volvo trucks (3 and 4 axles). The future transportation of coal from the mine will be done by 70 tones Volvo trucks or 60 tones semi-trucks. We have very good experience with "zeipa-tires" on the trucks.

We expect to spend approximately 5 NOK/ton on maintenance of the road this year. The total transport cost of coal to the harbor, including loading seems to be 2 NOK/ton & km (25 NOK/ton) over 12 km distance in addition to 5 kr/ton in loading to ship costs.

Mining in the Arctic, Olsen, Lorentzen & Rendal (eds), © 2001 Swets & Zeitlinger, Lisse, ISBN 90 5809 177 5

Waste disposal at the apatit mining company's phosphate operations, Kola peninsula, Russia

N. N. Melnikov, E. B. Krasnoselskiy & S. P. Reshetnyak
Mining Institute, Kola Science Centre, Russian Academy of Sciences, Apatity, Russia

ABSTRACT: Apatit Mining Company's open pits are taken as an example here to show the relationship between large scale waste rock dumping on mountain terrain and polar climate with a lot of snow. The natural conditions were investigated and waste dump deformations are described. Safe and cost effective technology was developed and adapted for Apatit Mining Company's phosphate operations at Khibiny Mountains, Kola Peninsula, Northwest Russia. Advanced dump disposal and operation under the severe northern condition saved a significant profit for Joint-Stock Company (now) "Apatit". This technology is acceptable for many other open pits located in mountains and with snowy winter.

The climate of the Kola area is milder than the same territories located to the East of the Kola Peninsula. The warm waters of the Gulf Stream warm the region and help keep the Barents Sea free of ice. Winter normally lasts 5 months. In the mountain regions (for example, in Khibiny) winter can last for 8 months. Along the coast the average temperature in January ranges between -9° C and -14° C. However, the fluctuations in temperature can range from -50° C to +10° C. The days with daily average temperature from -20° C to -30° C are rare. The average temperature in July is from +9° C to +14° C but it can vary from -3° C to +34° C sometimes. Humidity of 80% is normal 165 days per year. Average precipitation is 500 - 700 millimeter per square centimeter (mm/cm^2), in the mountains it is 900 - 1300 mm/cm^2, but the quantity of solid precipitation (snow) only for the winter period can be 900 - 1200 mm in Khibiny with a total yearly precipitation of 1500 - 1800 mm. Therefore, the snow depth can reach 3 - 4 m. The average snow cover is from 25 to 75 cm in the area, in the mountain region it is to 150 - 200 cm. In winter period the numbers of days with snowstorm range between 23 to 111 days with snow more likely in the mountain regions. Total snowstorm carrying during the winter in Khibiny can reach 1200 m^3/m of the front line. Sometimes in Khibiny you can see the unique phenomenon - the hoarfrost formation which results from the intrusion of moist northwestern winds from the Barents Sea (at air temperatures below -10° C). The hoarfrost thickness reached 240 cm (at a density 0.6 - 0.7 t/m^3) and the ice destroyed the supports of the air power line.

From 1964 the Centralny open pit of Apatit Company operates under these conditions in Khibiny. It is the largest open pit in the Arctic Circle. Centralny has produced to 28 million metric tons of apatite-nepheline ore per annum until 1990. Overburden and waste rock removing reached 14 million m^3 per year. The Centralny open pit is located on the Khibiny plateau named Rasvumchorr, 1050 m above sea level. It has steep mountain slopes in all directions, the height of the slope is 400 - 600 m. Waste dump failures are a problem as they can collapse suddenly with avalanche intensity.

Since early days the apatite-nepheline deposit development in the Khibiny Mountains has had to confront severe mining conditions in the Arctic environment. At first, it appeared due to heavy snow-fall, strong snow-storms, destructive avalanches and long winter time. Winters on the Kola Peninsula are milder than in central and northern Siberia but due to the quantity of solid precipitation, force and quantity of snow-storms Kola's winters do not compare with Siberian winters; especially it is characteristic for Khibiny mountain-range. Based on data from long-term records of Avalanche Protection Shop of Apatit Mining Company, the annual precipitation on the Rasvumchorr Plateau, where Centralny open pit is located, ranges from 1500 mm to 1800 mm of the water. Thus, 900 - 1200 mm is the normal winter precipitation.

Snowstorms are affected by the prevailing winds. In the vicinity of the ice-free Barents Sea and Khibiny

Mountains topography influences the wind. The wind velocity on the plateau tops can reach 60 meters per second (m/s) sometimes and as often as not reaches 40 m/s on Khibiny meteorology data. Winter winds generally have a northern western direction and by summer a southern western flow. Average snow carrying for winter on all directions reaches 1200 m^3/m.

Adduced glatioclimatical condition of Khibiny mountain range annually creates the avalanche dangerous situation during the winter. Only by the natural solid precipitation in Khibiny the snow thickness is to 3 - 4 m at snow density 0.33 t/m^3. Due to the snowstorms carrying to lower forms of relief, the thickness of snow can reach 5 - 6 m, and on convex sites the snow cover can partially be away level. Such snow accumulation on the mountain slopes can cause and ultimately result in occurrence of snow avalanches that can present a serious danger for mining operations.

Nevertheless, snowstorms and snow avalanches are not directly related to the mining operations especially on waste dumping, i.e. the corrective procedures are not available in the technology. The effective ways of preventive struggles both with snowstorm influence and with avalanches do exist. In present work these questions are not considered, as each of them is presented enough extensive and difficult to combine directly with the waste disposal. But glatioclimatical factors themselves (the precipitation, their quantities, the snowstorms, redistribution and below freezing temperatures) and their influence on the waste dump technology require a detail analysis.

Because of problems associated with complex mountainous terrain and adverse climatic condition, the waste dumps of the Apatit Mining Company have been the subjects of investigations by the Kola Mining Institute, Russian Academy of Sciences, since 1970. In particular, the research focuses on dump deformation, sliding, and caving, which pose hazardous risks and cause delay in the mining operations of the Centralny open pit mine. Both theoretical and experimental investigations were carried out in the laboratory and in the field.

Due to the government regulation, snow removal to the waste dumps was prohibited. The winter period lasts 8 months at the Centralny open pit. The period of active snow accumulation exceeds 7 months. Means that in this time the snow is accumulated in the waste dumps independently and even contrary to the will of man. Therefore, we have not only rock but a snow-rock mixture in the waste dumps. The snow-rock mix varies greatly: from 0% (snow free rock) to 100% (purely snow - vein or layer).

If the weather condition is not a tempest, dumping proceeds during the snowstorms. Works stops only

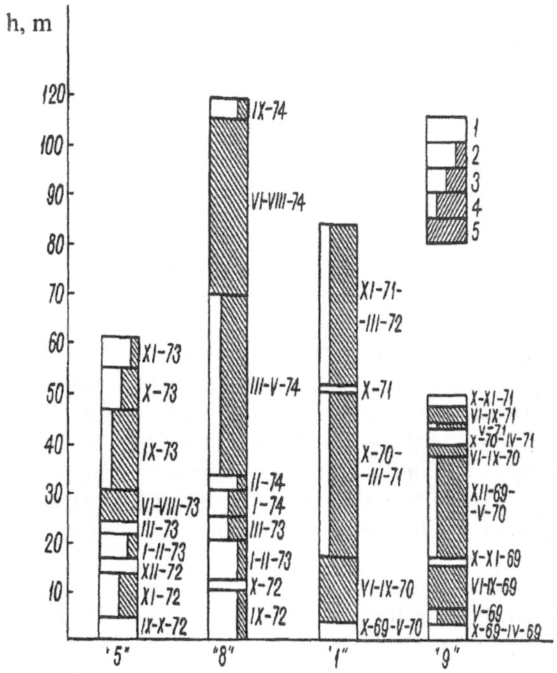

Fig. 1. Stratigraphic cross-section of waste dump at Centralny open pit: 1 - dense snow without rock; 2 - dense snow volume is more than pores volume of waste rock; 3 - dense snow volume is equal pores volume; 4 - snow volume is less than pores volume; 5 - pores is empty (no snow); “5”, “8”, “1”, “9” - number of dump

during severe snowstorms with low visibility. The fresh snow covers the waste dumps but also is collected in the waste rock during blasting. Long-term research has allowed for creating stratigraphic cross-sections for particular waste dumps (fig. 1). The stratigraphical section depends upon the rate of waste dumping and the accumulation of snow from snowfall and snowstorm.

Properties of snow-rock mix depend on the snow content of the waste rock and the volume of the snow buried in the dumps. The mix has five gradations (fig. 1). Contrary to the traditional opinion, snow in the waste embankment does not reduce it strength. On the its relative strength (tab. 1) increases temperatures' field for 2-nd, 3-rd and 4-th gradation (fig. 1) on 60-70%.

Thus snow in the waste rock can play the cementing role. However due to the specific properties of snow the strength of snow-rock mixture depends on its

Table 1. Dependence of shift resistance of snow-rock mixture from the snow content and the vertical load

Load, MPa	Depth of the point, m	Shift resistance (MPa) at the snow content (%)					Range of readings' increasing	
		0	5	10	15	20	MPa	%
0,3	14	0,28	0,33	0,38	0,42	0,45	0,17	61
0,5	23	0,42	0,52	0,57	0,65	0,72	0,3	71
0,7	32	0,6	0,73	0,84	0,93	1,01	0,41	68

Table 2. Dependence of shift resistance of snow-rock mixture from the temperature and the vertical load

Load, MPa	Depth of the point, m	Shift resistance (MPa) at the temperature of mixture (0° C)					Range of readings' increasing	
		0	5	10	15	20	MPa	%
0,3	14	0,28	0,33	0,38	0,42	0,45	0,17	61
0,5	23	0,42	0,52	0,57	0,65	0,72	0,3	71
0,7	32	0,6	0,73	0,84	0,93	1,01	0,41	68

temperature. The tab. 1 data are received at experiment in range of temperature from -3° C to -5° C. At 0° C the snow-rock mixture behaves like magma. The conducted research have shown that at snow content 10-20% the resistance to shift of the mixture decreases (tab. 2) by 35-45% with temperature increases from -2° C to -0.5° C.

Experimental research has shown that the heating of snow-rock mixture reduces its strength, especially - in range of temperatures from -2° C to 0° C. Such dynamics is characteristic of snow which comes nearer to point of the phase transition from solid to liquid condition. The force of coupling in the latter is less than in the first one considerably. Further, the strength of waste material depends on the degree of humidity. The snow-rock material is humidified to the utmost moment of the snow thawing. The loss of redundant water the waste dries out. It inevitably causes an increase in resistance to shift that is illustrated well by the experimental results (fig. 2).

According to data received, it is possible to draw some conclusions in waste dumping technology under the difficult glatio-climatic conditions in the mountains, i.e. under the conditions of snow accumulation and below freezing temperatures:

1. Under the difficult glatio-climatic conditions is not presently to avoid of the snow accumulation in the waste dumps at the mining operations around year. Snow falling on waste dumps is a natural process, due to snowstorms, as well as the dilution of waste with snow at the working face.

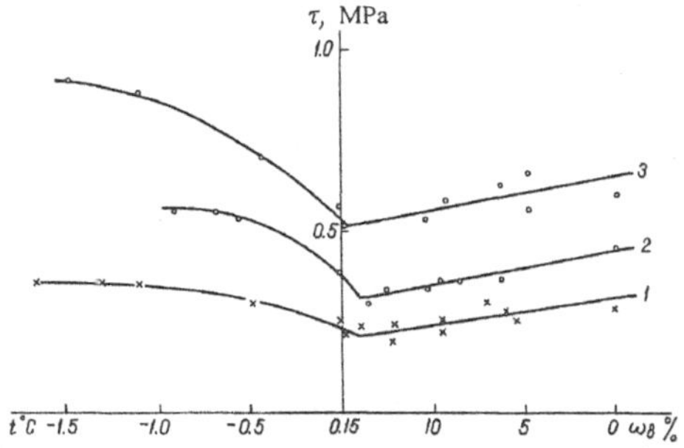

Fig. 2. Dynamics of shift resistance (τ, MPa) in snow-rock dump during the heating (t, °C), snow thawing, getting wet and drying up (ω_m, %) at different loading connected with various depth: 1 - 0.3 MPa, 14 m; 2 - 0.5 MPa, 23 m; 3 - 0.7 MPa, 32 m

2. Snow in the waste dumps at below freezing temperatures increases the strength of structural connections in waste dumps.

3. Reduction of structural strength connections in the snow-rock dump can happen with the heating and the thawing of snow. Two variants are thus probable:

a) snow is completely distributed in the pores of waste rock and does not break the skeleton of the waste dump. At the snow thaws, the strength of the dump is reduced to level of strength of the damp rock of appropriate humidity;

b) snow in the dump forms snow layers or lens. As the snow thaws, the strength of dump is reduced in general. The danger of possible slides forming in locations of existence lenses is increase. Hence, the caving of waste rock (prisms) is possible.

4. In anticipation of caving we can apply special rules for waste dumping in a winter period. With the purpose of elimination of prism collapses, it is necessary to avoid the occurrence of snow layers and lenses in the dumps.

5. Previous conclusions permit one to conclude that removal of inpit snow is mandatory (from roads, berms, benches and faces) to acting waste dumps at indispensable completion of definite rules appropriate mixture of rock and snow.

Technological decisions were made for mountain waste dumping under the Arctic condition using long-term research of the thermo-cryogenic conditions of stockpiling of large rock volumes on the mountain slopes of Khibiny:

1. Average temperature of waste rock layers formed in various months winter season approximately corresponds with the monthly average air temperature.
2. As a result of winter dumping of the frozen snow-rock, core is created according to distribution of below freezing temperatures between layers in the waste dump body.
3. Due to internal heat exchange between various layers of winter and summer dumping, below freezing temperatures are established inside the dump ranging from -5° C to -3° C.
4. During the summer period, the zero isotherms penetrate to a depth of 2-3 m below the surface. Increasing of below freezing temperatures can reach 5-10 m in depth.
5. Waste dump heating of the mountain slope surface is distributed to 2-3 m below the surface in the absence of thawing or underground water flows. Practically it does not destroy the waste dump stability.
6. Heating of the snow-rock waste dumps is possible only in the case of thawing or underground water flows inside waste dump are powerful heat-carrier capable of melting the snow buried inside the dump; the zero isotherm decreases to a depth of 10-15 m and more. With the thawing of the dump, instability within the dump can cause slides.

Thus, research reveals the true role of snow in a waste dump. Moreover a series of engineering measures can prevent failure by preserving the cementing role of the snow. The viscous-plastic properties of the snow-rock material, when to taken into account, can be used to stress relief the formation minimizing failure.

Regulations of waste dump construction under normal snow accumulation provided lower insitu temperatures providing a more stable structure. For this purpose, rock dumping is proceeded by front stopes with definite length and with definite step of progress. Thus, the dumping intensity should be such that the pore volume completely contained the volume of accumulating snow. The volume snow accumulation expected is determined according to average long-term meteorological data for the solid precipitation and the snowstorm carrying.

A regulation of the snow-rock removed from the open pit by is based on the principle of total accommodation of snow stewing to waste dump in pores of the dump skeleton. The ratio of snow and rock mixing is carefully monitored. This ratio can change by specified criteria during the snow fall period.

Thinking of the mountain waste dump formation as kind of a snow-rock mass exclude the unexpected discontinuities in their steady can mitigate sudden caving. However, the snow availability increases deformation ability of the mountain waste dumps. The deformation ability of snow-rock dump has a monotonous and regular nature. Thus, two kinds of deformation were allocated: the deformation of condensation which accompanies any discrete material dumping is connected with its condensation and tends to attenuation, and second kind - the deformation of sliding which arises and develops with growth of the load in the dump and its parameters associated with its ultimate size. Beyond these parameters caving will occur. Controlled deformation can be acceptable engineering solution. The deformation velocity under controlled criterion provides for safe working conditions on the dumps.

Condensation tends to increase sliding grows in accordance with the waste dump weight relative height. The general waste dump deformation as a result of its development on the mountain slope inevitably results in it collapse. Analysis of deformation development of several dumps before their collapse has shown (fig. 3) that the dynamics of deformations is subject to its general regularity. At a deformation rate of 50 cm per day, a dump is considered non-dangerous. The nature of increasing deformation velocity is monotonous and sudden dump caving does not take place. At a deformation rate of more than 50 cm per day, a dump can collapse very quickly. Because the step deformation of surface of the operating dump zone happens so fast, a large part of working day has to be spending

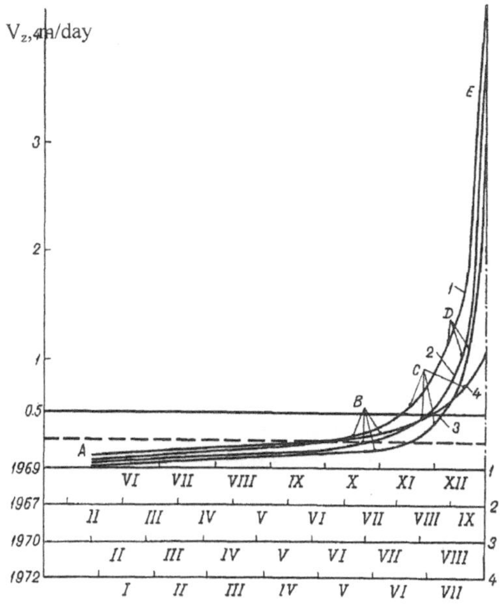

Fig. 3. Change of waste dump deformation rate (v, m/d) at continues increasing its volume up to caving (right axis is the caving time): AB - condensation is main; BD - sliding is developing intensively; DE - sliding is main, process is going to caving irretrievably; C - point of criterion to work safely on the waste dump, sliding begins to exceed condensation

controlling dumping of rock smoothing the surface. Obviously, such control of dumping is not economically feasible. Therefore, at a deformation rate of 50 cm per day, it is necessary to terminate dumping. This deformation rate is not critical for a dump but it serves as a safety criterion.

Deformation on an active dump, as a rule, has the maximal unloading at the waste dump edge. This regularity is established at supervision level incorporated in directing trucks to the rock base edge. On each profile line from 3 to 7 reference marks were established dependently from the dump width (fig. 4). The dependence line between the deformation rate, ***v***, of the embankment part of operating dump surface and the distance, ***r***, from the mountain base of the dump is has been shown that for an increasing value of *r*, *v* increases exponentially. *V* has a maximum value at the furthest part of dump. The touch line is the limits of the trust interval with probability 95%. The bench form is in according with the nature of the bank part of the failing dump surface. This peculiarity was marked practically on all high waste dumps in Khibiny.

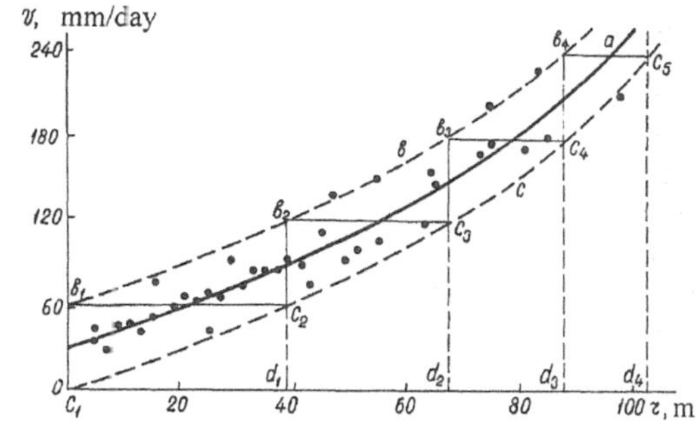

Fig. 4. Dependence of dump berm deformation rate ***v*** from the distance ***r*** between the point of measure and the rock base edge: a - deformation rate; b, c - trust range internal limits

Regulation of survey controls for the deformation condition of mountain dumps is located at the base the snow-rock dumps during their formations. Supervision on the working reference marks in unloading zones and on the working perimeter of dump has reduced the volume of survey work on 70-80% and preserved the safety.

Ultimate width of the operating mountain dump under steady condition is the distance from the rock edge to the dumping edge. It is entered as the new criterion and the new parameter similar ultimate height of steady dump. Such replacement is produced for the reason that the current height of mountain dump on the slope does not allow survey control for reason of constant moving of the dump bottom edge and its not correct definition. Operating dump width is easily controlled by monitoring the approach - needed for waste dump stability reliability.

Maximum height is another criterion of dump condition on the high mountain slope. It serves for propose the stable dump condition. The stable dump condition depends on the physical-mechanical properties of waste rock and on the angle of inclination of mountain slope surface on which the dump is forming. The settlement formula for ultimate dump height is given as:

$$H = \frac{2\tau_n}{\gamma} \cdot \frac{\sin 2\alpha \cdot \sin(\omega_n - \beta)}{\sin(\alpha - \beta)\left[\dfrac{(1 - \sin\varphi_n)\cdot\sin(\alpha - \beta)}{\sin(\omega_n + \beta)} - 2\cos\beta \cdot \sin(\omega_n - \alpha)\dfrac{\sin(\varphi_n - \beta)}{\cos\varphi_n}\right]},$$

where φ - internal friction angle; ***n*** - reserve ratio; β - top half of slope angle under the waste dump; τ - minimum shift resistance of waste, t/m^2; γ - medium density of waste, t/m^3; α - slope dump angle; $\omega_n = \text{arctg}[(\text{tg}(45+\varphi/2)/\boldsymbol{n}]$.

For the value of the n-factor of stability, it is necessary to say more in detail development and dynamic effects arising during the dump construction:

n_1 = 1,07 - takes into account insufficient reliability of rock samples used for definition of physical-mechanical characteristics; n_2 = 1,2-1,3 - takes into account the error of crack property account of rock and semi-rock waste, as well as at the expense of decrease of shift resistance due its creep ability; n_3 = 1,15-1,2 - takes into account the errors of definitions of average density of rock and methods of account as well as the error from dynamic load at blasting of large explosive charges; n_4 = 1,2-1,3 - takes into account the seismic load from effects of earthquakes of magnitude 6 - 9 (on Richter's scale).

At simultaneous accounting for several factors, the general reserve (safety) factor is equal to multiplication of separate reserve factors: $n = n_1 \times n_2 \times n_3 \times n_4$. The reserve factors can be accepted in dependence on particular natural and mine technical conditions as well as - in dependence on time. In this case, the value of the factor is accepted in lesser half of the range of significance if the settlement time is to be 5 years, and in greater half - if it exceeds 5 years.

Obviously, that the use of this formula is rather inconvenient and is labor consuming. On base of this equation, the evaluation of height limit of stable dump is developed in nomogram form that permits fast definition of the required parameters.

The nomogram for finding an ultimate width was made to simplify usage. On computer base the nomogram permits to determinate the dump width using the slope angle below the waste dump and average minimum values of shift resistance. Comparison of the height limit of a stable dump and its width limit has been shown to be practical. The analytical formulas of their mutual calculation are received.

Block-level dumps as a new construction tool provides an obviously stable waste dump on high slopes. They are developed on the basis of valuation of limit parameters of snow-rock dumps and sequence of dump formation by stable blocks and levels. This way of designing dumps preserves the strewing front as a stable, expedient and convenient means for assuring long-term stability.

Waste discharging over the bank generated from snow-rock material is based on a safety factor on the surface of the most probable failure of the bank with a loaded truck. The safety factor depends on ratios of parameters of bank and truck capacity and is not less than five-divisible size of force acting on bank in the break plane. This way of unloading on the dump instead of the waste unloading on the dump surface with subsequent moving on the dump slope by dozer considerably has reduced and made less expensive (3 - 5 time) of the waste dump formation.

The technological decisions considered on mountain waste dumping are developed on the basis of the long-term research and experience of dump operation in the Khibiny's open pits. Each technological decision was tested and introduced in the waste dump practice. It produced safer, less costly waste dumping under difficult terrain and glatioclimatical. The set of given technological decisions presents the technology of mountain waste dumping adapted to severe Arctic conditions and difficult relief.

The results of investigations (more than 20 developments) were introduced to the mining operation. Major developments have been realized in the following areas:

1. The structuring or ordering of waste dump formation on steep rock slopes.
2. Design and control of mine surveying to improve safety on constantly sliding dumps.
3. The transport, structuring, and ordering of combined waste and snow dumps transported from the pit.
4. Systems control of moving dumps to enhance and maintain a stable waste dump disposal front.
5. Development of formation principles that consider the self-transporting characteristics of waste along the slope.
6. The structure and formation of stable block-level foundation dumps of large capacity.

The technological factors listed do not resolve all the complex technological decisions connected with waste dumping in mountain terrain. Anyone having interest in this information can obtain it on a case-by-case basis.

Mining in the Arctic, Olsen, Lorentzen & Rendal (eds), © 2001 Swets & Zeitlinger, Lisse, ISBN 90 5809 177 5

Centrifuge physical modelling of frozen backfill

E. De Souza & A. P. Dirige
Department of Mining Engineering, Queen's University, Canada

ABSTRACT: This paper presents a unique frozen backfill design investigation based on centrifuge physical modelling and static strength tests. The study aimed at establishing the most suitable frozen backfill design which would provide superior ground support during mining, or during pillar recovery, with minimum dilution and minimum water run off. The design study considered the material properties, and temperature and saturation effects. The strength and behaviour characteristics of the engineered backfill designs have been established to permit implementation in the field.

1 INTRODUCTION

Underground mining operations in the Canadian Arctic operate in permafrost ground conditions. Although frozen rock exhibits superior strength and demonstrates low susceptibility to ground disturbance (Kelly 1983), mining development may result in thawing of the rock mass due to heat transfer, with subsequent reduction in rock strength. As the thaw front advances into joints and fractures in the rock mass, reduction in rock strength would result, thus compromising the stability of the workings. Underground excavations need to be backfilled to guarantee stability under such unique rock conditions. The introduction of a suitably designed frozen backfill into mined out stopes would prevent the deleterious weakening of the rock and minimize ore dilution.

This paper presents a unique frozen backfill design program based on centrifuge physical modelling. The design investigation considered two candidate materials, limestone and siltstone, quarried at a hard rock mine operating in the Canadian Arctic. Frozen fill scaled models were built with different water contents, and cured at different temperatures and times. Centrifuge modelling permitted accurate qualification and quantification of the strength and behaviour characteristics of the different frozen fill designs. Model results were corroborated with uniaxial compressive strength test data.

Before results of the frozen fill design study are presented, some basic concepts of the mechanics of frozen materials are introduced.

2 THE MECHANICS OF FROZEN MATERIALS

2.1 *Properties and behaviour of ice*

The unconfined compressive strength of polycrystalline ice is approximately 1 MPa (Andersland & Anderson 1978, Sprott 1983). The Young's modulus of ice is dependent on temperature and on the rate of strain it is subjected to. Polycrystalline ice has a Young's modulus ranging between 0.0569T and 0.0648T (GPa), where T is the temperature. However, for very low strain rates typical of underground excavations (10^{-7} s^{-1} or lower), polycrystalline ice will have a much lower modulus of approximately 4500 MPa. The Poisson's ratio of ice increases with temperature and normally ranges between 0.31 and 0.55. Ice may behave as a brittle (elastic) or viscous (time-dependent) material depending on conditions of temperature and strain rate. Creep rates increase with increases in temperature, however creep can be controlled by confining the ice.

2.2 *Behaviour of frozen backfill*

The strength of frozen materials develops from cohesion, interparticle friction and particle interlocking and is dependent on particle size (Andersland & Anderson 1978). Cohesive forces result from adhesion between backfill particles and from ice bond in the backfill voids. Interparticle friction, which is maximum in the stages of plastic deformation, greatly affects the backfill's peak and long term strength. A well graded particle distribution will produce a high density backfill, within which interparticle void space is reduced. Ice formed in such smaller voids will possess increased strength due to improved crystalline structure (Kelly 1983).

The mechanical behaviour of the frozen fill clearly depends on the water content. Lower saturation provides insufficient water to form an ice cementing matrix; supersaturation creates an ice matrix with insufficient particle contact to control excess ice creep. At low water contents interparticle friction thus determines the strength characteristics of the frozen fill while at very high water contents, the strength properties are similar to those of ice. For example, testing with typical sand mixtures has indicated that for low volume contents of sand, less than 40% sand by volume, the increase in strength is low (Goughnour & Andersland, 1968, Tsytovick 1975). When a threshold volume content of sand, about 42%, is reached, a rapid increase in strength is noted as friction between sand particles and dilatancy begin to play a role to the strength of the frozen sand.

Fill strength in also dependent on temperature, strain rate, particle size, and compaction. Elevated temperature results in non-ideal crystal development of the ice matrix and consequent reduction of the fill strength (Kelly 1983). Fill strength thus increases with decreases in temperature and the Young's modulus decreases with decreasing water content. As thawing of the backfill could quickly deteriorate ground conditions, heat sources must be carefully controlled and underground temperatures must be maintained at appropriate levels. Some of the primary sources of heat in a mine include ventilation air, the heat generated by mining equipment, the heat released by workers, and the heat released during blasting.

The creep rate of frozen fill is dependent on stress, temperature and composition. The creep rate increases with increasing water content. The creep strength of frozen backfill is the stress level at which rupture or extreme large deformations, about 20%, without rupture, occurs (Andersland & Anderson 1978). Backfill low in ice content typically rupture at critical loading while ice-rich backfill deforms plastically at similar stress levels. Excess water in the backfill will cause it to behave as ice, and excessive creep will occur even when the frozen fill is subject to minor loadings. The failure behaviour of frozen backfill of proper density and saturation levels would thus be plastic; it most likely would deform while failing rather than being susceptible to rupture.

3 MATERIAL PROPERTIES OF THE TESTED FROZEN BACKFILL

The tested materials originated from an operating hard rock mine in the Canadian Arctic. Two different fill materials are quarried for backfill production, limestone and bituminous siltstone.

The received limestone material had a moisture content approximating 5-7% and the siltstone material a moisture content of 4-5%. Measured bulk densities for the limestone and siltstone were 1308 kg/m^3 and 1367 kg/m^3, respectively.

The particle size analysis of the aggregate materials is shown in Figure 1. The particles sized greater than 25.4 mm were sieved out and only the portion with particle size less than 25.4 mm was used for the test specimens and models in order to minimize size effects. According to the grain size distribution the fill materials can be classified as sandy gravel. It is noted that the process of dumping fill into boreholes to fill the stopes will produce a finer distribution in the stope than the shown in Figure 1. As the frozen fill strength is dependent on the fines fraction, the size distribution at the feed has to be controlled at the mine to achieve the desired distribution in the stope.

The shape of the particles of solids making up the fill plays an important role on the strength of frozen fill as it develops from interparticle friction and particle interlocking. Grain shape is defined by a measure of the sharpness of the corners (Archibald & Nantel 1986). Both materials, as received directly from the mine quarries, exhibit sharp corners, and would be expected to greatly influence the frozen backfill strength. It is noted that the final grain shape of the fill would be the result of transportation to the disposal area and of dumping into boreholes to fill the stopes.

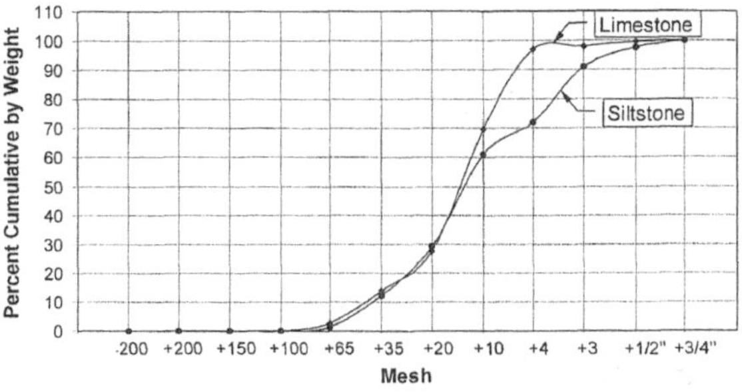

Figure 1. Size distribution curves

4 COMPRESSIVE STRENGTH TEST PROGRAM

The received materials were weighed and mixed with the required amounts of water and placed and tamped in layers into 102 mm by 203 mm plastic molds to give a specimen length to diameter ratio of 2:1. Two water:solids ratios by weight were used: 5:95 and 10:90, equivalent to 5% and 10% by weight water content. Placement densities averaged 1769 kg/m^3 and 1928 kg/m^3 for the limestone fill at 5% and 10% water content, respectively. For the siltstone fill, prepared at 5% and 10% water content, placement densities averaged 2039 kg/m^3 and 2011 kg/m^3, respectively. A T type thermocouple was placed in the centre of each specimen during preparation. The specimens were placed in a freezing unit, protected from moisture loss and from sublimation effects, and monitored for temperature. Four temperatures were used for the tests, 0°C, -1.2°C, -4°C and -12.9°C, consistent with expected fill temperatures underground.

Once the core of the specimen reached the test temperature, the plastic mold was removed and the specimen tested immediately on a Materials Testing System servo-controlled compression loading frame, at a constant stroke rate of 0.2857 mm/min, corresponding to a strain rate of 2.346 x 10^{-5}/sec. Figure 2 shows pictures of a frozen fill specimen immediately prior to and during a typical test.

Typical specimen response curves are shown in Figures 3, 4, 5 and 6 and a summary of results is given in Table 1. Test data clearly indicates the increase in strength with decreasing temperature and with increasing water content. The different frozen backfill recipes exhibit remarkable strength properties, in particular at the lower cured temperatures. For comparison purposes, the frozen fill uniaxial compressive strengths are consistently higher than that of non-frozen cemented sand backfills, typically ranging between 0.48 and 0.68 MPa, at 5% cement. The frozen fill strength data permits selection of specific recipes for particular applications underground.

It is interesting to note that the siltstone fill presented higher strength than the limestone fill at

Figure 2. Frozen fill specimen prior to and during compressive strength testing

corresponding water content and temperature. As will be discussed later, the siltstone frozen fill demonstrated inferior performance that the limestone fill when dynamically tested in the centrifuge.

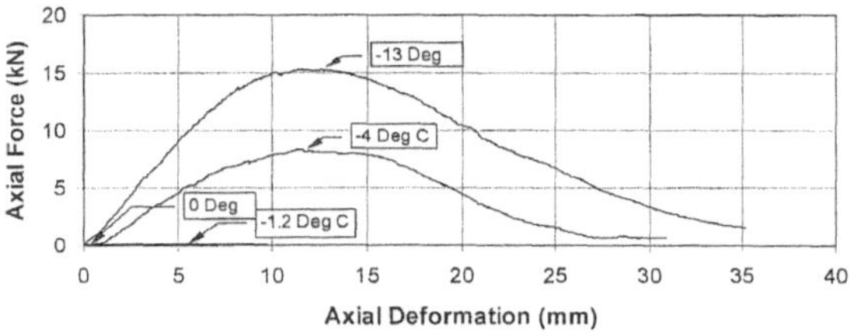

Figure 3. Load-deformation curves. Frozen limestone fill at 5% water

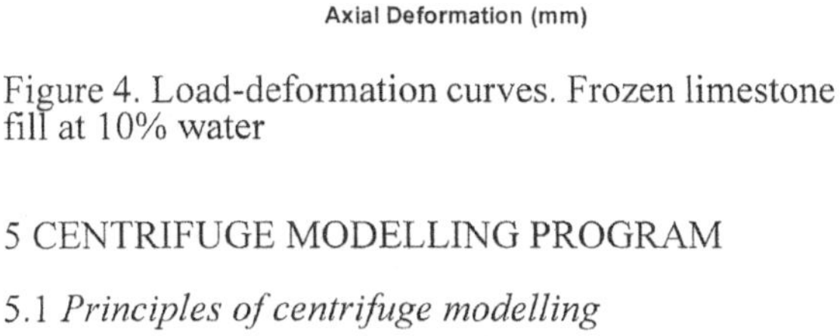

Figure 4. Load-deformation curves. Frozen limestone fill at 10% water

5 CENTRIFUGE MODELLING PROGRAM

5.1 *Principles of centrifuge modelling*

Modelling is one of the most important engineering tools used for design. In mining engineering, empirical, physical and numerical modelling methods are commonly used. Despite the current strong tendency towards computerization, rock engineering design still relies on experimentation as more complex theories require more testing. The principal benefit of developing scaled models is that problems that are too complex to be treated analytically or numerically can be effectively studied in the centrifuge. This is particularly true for materials of complex behaviour such as frozen fills. Also, scaled models permit the transformation of unmanageable properties of the prototype such as extreme size, very slow flow,

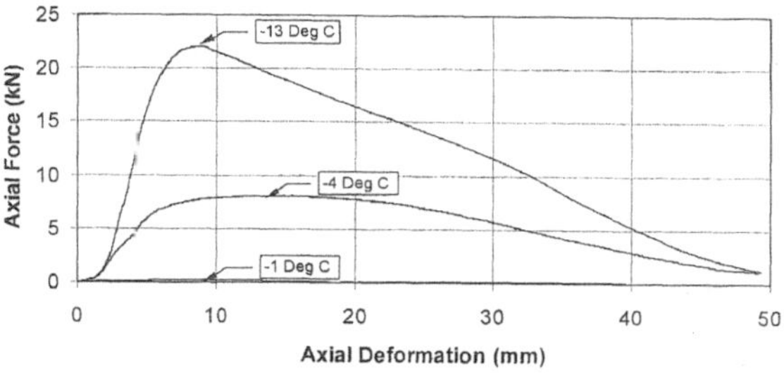

Figure 5. Load-deformation curves. Frozen siltstone fill at 5% water

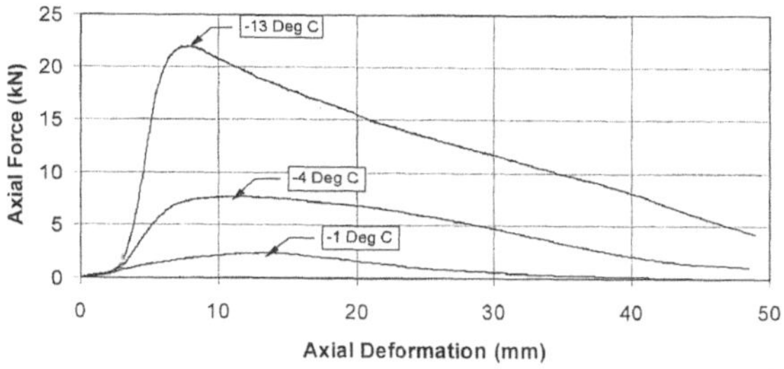

Figure 6. Load-deformation curves. Frozen siltstone fill at 10% water.

Table 1. Uniaxial compressive strength of frozen fill

Fill Recipe (L: limestone S: siltstone)	Placement Density (kg/m^3)	Peak Strength (MPa)	Modulus (MPa)
L, 5% w.c., 0°C	1806	0.002	-
L, 5% w.c., -1.2°C	1732	0.017	0.439
L., 5% w.c., -4°C	1760	1.042	14.605
L., 5% w.c., -12.9°C	1778	1.957	46.664
L, 10% w.c., 0°C	1958	0.029	1.783
L, 10% w.c., -1.2°C	1927	0.054	0.503
L, 10% w.c., -4°C	1922	1.275	19.523
L, 10% w.c., -12.9°C	1905	3.106	124.50
S., 5% w.c., 0°C	2029	0.031	0.66
S, 5% w.c., -4°C	2023	1.017	37.91
S, 5% w.c., -12.9°C	2065	2.794	126.09
S, 10% w.c., 0°C	2001	0.301	6.96
S, 10% w.c., -4°C	2021	0.969	44.22
S, 10% w.c., -12.9°C	2010	2.785	179.80

extreme energy release and microscopic dimensions to manageable domains. Because the centrifuge is used to simulate gravitational forces on a model, it duplicates the mass stresses that create the phenomenon of deformation or flow. As a result, the lengths, volumes, forces, stresses, velocities, accelerations, flows and time of the prototype, are scaled such that the model can be used as an adequate substitute for the original. If scaled correctly by observing the conditions of similitude, patterns of deflections, deformations, distortions, arching, deflection, cracking, failure and a multitude of other related properties measured on the scaled model permit prediction of the corresponding properties in the prototype design. Scaled models are cost effective and represent an effective alternative to full scale tests; model results from a centrifuge can provide reliable and consistent data for viable engineering designs.

Centrifuge modelling is based on the fundamental concept that gravitational mass forces can be simulated by centripetal mass forces applied by the acceleration of a scaled model. The stress levels acting within an N-scaled model accelerated at N times the earth's gravity field are analogous to those acting on the prototype. For centrifuge modelling to be valid, similitude must exist between the accelerated scaled model and the prototype. Conditions of similitude are satisfied by using two approaches: by applying scaling laws and by performing dimensionless analysis.

Regardless of the type of application, the principles and design of scale model experiments are similar and have been well established. First, a geometrically-similar model, scaled down at 1/N, is formed with materials of identical properties as that of a prototype. The transformation of the prototype must ensure that the model will have the same relative position, proportions, values and structure. The model is, in essence, a small prototype. The scale model is then placed in the centrifuge strongbox and rotated at a desired angular velocity. A corresponding centripetal force acceleration is experienced by the model in flight, and this acceleration induces self-weight forces in the test material. If a 1/N scaled model is accelerated to N gravities, the inertial mass induced stresses at any point in the model will be identical to the gravitational mass induced stresses at any similar point in the prototype, providing that the boundary conditions are identical at all equivalent points (Mitchell 1991). All stress-induced and stress-gradient induced events will, then, occur in the model as they would in the prototype (Mitchell 1991), but model events depending on gradients will occur faster because the physical distances are reduced in the model. Thus, the model accelerated in the centrifuge experiences analogous behaviour as the prototype.

5.2 *The centrifuge testing facility*

The centrifuge is a 5.3 m diameter machine driven by a 45 kW hydraulic motor and speed control assembly. It is designed to rotate up to a maximum speed of 350 rpm and it is capable of exerting up to 300 gravities on typical models. It is rated at 30g-tonne, since it can accelerate a 0.1 tonne mass to 300 gravities.

The centrifuge is located in a below-ground room, and is enclosed by a 0.15 m thick, 5.3 m diameter, 1.4 m high, circular reinforced concrete wall. The centrifuge pit serves as a protective enclosure for the facility. A cantilevered reinforced concrete ledge, 76.2 cm wide and 10.16 cm thick, extends over the strongboxes, and is used to provide containment for the system. The control room is adjacent to the centrifuge room, and is shielded by a concrete block wall to provide for additional safety.

The main beam of the centrifuge is 3.85 m long and is fabricated from a 20 cm steel box section of 1.3 cm wall thickness. An aluminum swing strong box, 0.75 m long, 0.56 m wide and 0.42 m high, attached to one end of the main beam by a 7 cm diameter high strength steel pin, is used to house scaled models. A steel counter-weight box, 0.71 m long and of 25.4 cm square section, is attached to the other end of the main beam by two 3.5 cm diameter steel pins. The beam rotates on a 7.6 cm diameter steel shaft which is powered by the 45 kW hydraulic motor. The system is belt driven, on a 1:1 ratio, using two 14M, 68 tooth polychain drive pulleys and a 60 mm wide belt. A 56 kW, 0.005 m^3/s, 10.34 MPa, 550 VAC 3 phase, hydraulic pump provides power for the system. The speed is controlled by ramping mechanisms in the hydraulic power supply, and can be maintained at any level between 5 and 350 rpm.

In centrifuge physical modelling, instrumentation and video monitoring represent the primary means of studying the behaviour of scaled models during testing. This is because centrifuge modelling is qualified by physical measurements and by observation. A high speed data acquisition and control unit is used for amplifying, multiplexing and digitizing analog or pulse signals from various sensors utilized to monitor the behaviour of scaled models. Such signals are transmitted to the data acquisition system through electrical slip rings assembled at the centrifuge shaft. Digitized signals are then processed and stored in a computer. The centrifuge has a total of 14 slip rings. Models can be viewed and video taped during flight through a port in the centrifuge concrete pit, and with the aid of a high intensity strobe light. Centrifuge monitoring and diagnosis are performed by a video surveillance system and by proximity sensors.

5.3 *Model preparation and modelling procedure*

Models were designed to represent a prototype stope, planned to be backfilled with frozen fill of low water content in order to provide support to adjacent frozen ground and to cemented rock fill to be exposed by mining. Of primary concern is the possibility of heat transfer to the frozen fill, resulting in excessive run off water and in associated problems such as the need to build dams and pump stations, water freezing in drill holes and the desensitization of explosives.

The prototype stope is 15 m wide, 40 m long and 27 m high up dip and 18 m high down dip. The corresponding centrifuge models were designed at 75 g, having scaled dimensions of 19 cm wide, 55 cm long and 37 cm high up dip and 24 cm high down dip. The stope walls were vertical and boundary conditions were represented by smooth walls. Figure 7 shows representations of the prototype stope and centrifuge model.

A total of six models were prepared, four models were constructed using the limestone material and two models using the siltstone material. Treated form ply was used to construct curing molds, built to the design scale for the models. The received materials were

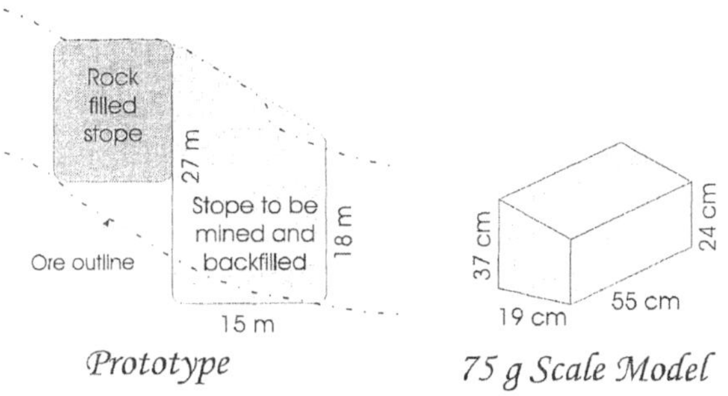

Figure 7. Prototype and model representations

weighed and mixed with the required amounts of water and placed and tamped in layers into the curing molds. Two water:solids ratios by weight were used, 5:95 and 10:90, corresponding to 5% and 10% by weight water content. The required weight of limestone fill per model approximated 35 kg and 37 kg at 5% and 10% water content, respectively. Required material quantities for the siltstone fill approximated 39 kg and 40 kg at 5% and 10% water content, respectively. Placement densities averaged 1080 kg/m^3 and 1172 kg/m^3 for the limestone fill at 5% and 10% water content, respectively. For the siltstone fill, prepared at 5% and 10% water content, placement densities were 1260 kg/m^3 and 1224 kg/m^3, respectively. T type thermocouples were placed in the centre and close to the face of each fill model. The curing molds were placed in a freezing unit, protected from moisture loss and from sublimation effects, and monitored for temperature. A range of temperatures were used for the tests, consistent with expected fill temperatures underground. For the limestone fill, cure temperatures were 0°C, -0.3°C, -1.2°C, -4°C and -12.9°C. Cure temperatures for the siltstone fill were -0.7°C and -0.8°C. Table 2 presents the model design data. The cure time represents the time after the core of the fill reached the required temperature.

Once the core of the model reached the test temperature and the required cure time had elapsed, the fill was exposed and painted with stripes to permit visual observation of behaviour - deformation, consolidation, fracturing, etc. - during flight. The cured frozen backfill specimen was then placed in a model box, specially constructed to accept the frozen specimens. The model had its walls insulated to minimize temperature variations during the test runs. The profile of the frozen fill stope could be seen via a 12.7 mm thick Plexiglass sheet secured on top of the model box. The model box was secured in the centrifuge strongbox and the test proceeded immediately to preclude the frozen fill from melting. Model weights, which constituted the strongbox carrying payload, ranged between 46 kg and 52 kg. A typical model ready for testing is shown in Figure 8.

The first stage of the centrifuge modelling procedure was to condition the model for 15 minutes at a centrifuge speed of 60 rpm, corresponding to an acceleration of 10 g. The speed was then increased at a rate of approximately 3 rpm per minute, until either the fill column failed or the design scale of 75 g was achieved. Observations of fill behaviour were recorded during flight and the models were photographed for further assessment.

Table 2. Centrifuge model design data

Model	Fill Recipe	Placem. Density (kg/m^3)	Cure Time (days)	Model Weight (kg)
1	L, 5% w.c., -0.3°C	1093	9	34.88
2	L, 10% w.c., 0°C	1186	14	37.85
3A	L, 5% w.c., -12.9°C	1066	4	34
3B	L, 5% w.c., -4°C	1066	4	34
3C	L, 5% w.c., -0.3°C	1066	5	34
4A	L, 10% w.c., -1.2°C	1158	4	36.94
4B	L, 10% w.c., -0.2°C	1158	5	36.94
5	S, 5% w.c., -0.8°C	1260	0	40.19
6	S, 10% w.c., -0.7°C	1224	0	39.05

Figure 8. Model in strongbox ready for testing

5.4 *Centrifuge model results*

A total of nine centrifuge model runs were performed, six runs using different limestone frozen fill recipes and two runs using the prepared siltstone frozen fill models. Table 3 presents a summary of fill model performance results. All models were scaled at 75 g, to represent a backfilled stope 15 m wide, 40 m long and 27 m high up dip and 18 m high down dip.

Model 1 represented a stope backfilled with frozen limestone fill, placed at 5% water content and cured at a temperature of -0.3°C. The frozen fill mass did not fail when loaded to 79 g. Minor sloughing, approximating 1-2% of the total fill volume, occurred at the free face. Consolidation of the fill mass was in the order of 0.6 cm, or 0.5 m at 75 g. Figure 9 shows model 1 at the end of testing. This simulation indicates that limestone frozen fill placed at low water content,

Table 3. Centrifuge model results

Model	Fill Recipe	Centrifuge rpm at End of Test	Gravity Loading at End of Test	Behavioural Information
1	Limestone, 5% water, -0.3°C	172	79 g	no failure 1-2%vol. sloughing 0.5 m consolidation
2	Limestone, 10% water, 0°C	170	77 g	15%vol. slough failure 0.5 m consolidation
3A	Limestone, 5% water, -12.9°C	169	76 g	no failure no consolidation
3B	Limestone, 5% water, -4°C	170	77 g	no failure 1%vol. sloughing 0.5 m consolidation
3C	Limestone, 5% water, -0.3°C	170	77 g	10%vol. slough failure 1.5 m consolidation
4A	Limestone, 10% water, -1.2°C	169	76 g	no failure 1-2%vol. sloughing 1.5 m consolidation
4B	Limestone, 10% water, -0.2°C	169	76 g	no failure 5%vol. sloughing 1.5 m consolidation
5	Siltstone, 5% water, -0.8°C	54	8 g	20%vol. slough failure
6	Siltstone, 10% water, -0.7°C	45	5 g	25%vol. bulk failure

Figure 9. Limestone fill (5% water, -0.3°C), model 1

even at temperatures reaching the freezing point, still retains structural strength.

Model 2, prepared at 10% water content and cured at a temperature of 0°C, experienced failure when loaded to 77 g, represented by collapse of approximately 15% of the fill material close to the free face. The centre of the fill mass remained intact, with no signs of fracturing or structural deterioration. Consolidation of the fill mass was in the order of 0.6 cm, or 0.5 m at 75 g. Figure 10 shows model 2 at the end of testing. This simulation indicates that limestone frozen fill at higher water content may experience

Figure 10. Limestone fill (10% water, 0°C), model 2

substantial sloughing at the face when temperatures reach the freezing point.

Model 3A was prepared at 5% water content and cured at a temperature of -12.9°C. The frozen fill column did not fail when loaded to 76 g; no wall sloughing and no consolidation of the fill mass were either noted during the dynamic centrifuge loading process. The model temperature was then increased to -4°C and the frozen fill model was re-tested as model 3B. During loading, minor sloughing, approximating 1% of the total fill volume, occurred and consolidation of the fill mass reached 0.6 cm, or 0.5 m at 75 g. Loading reached 77 g with no failure of the placed frozen fill. A further increase in temperature to -0.3°C was permitted and model 3C was loaded to 77 g, where approximately 10% of the backfill mass failed, primarily close to the free face. The core of the fill, at the centre of the stope, was still frozen and intact at the end of the test. Consolidation further increased to 2 cm, or 1.5 m at 75 g. This stepped testing approach at increased fill temperature attempted to establish the fill response to underground conditions of heat transfer to the frozen fill mass. Figure 11 shows model 3C at the end of testing.

Figure 11. Limestone fill (5% water, -0.3°C), model 3C

Model 4A, prepared at 10% water content and cured at a temperature of -1.2°C did not fail when loaded to 76 g. Minor sloughing, approximating 1-2% of the total fill volume, occurred at the free face. Consolidation of the fill mass was in the order of 2 cm, or 1.5 m at 75 g. The model temperature was permitted to increase to -0.2°C and was re-tested as model 4B. No further consolidation and no fill failure was noted during loading to 76 g, although sloughing at the fill face increased to 5% of the total fill volume. Figure 12 shows model 4B at the end of testing.

All frozen fill modelled stopes using the limestone material have demonstrated superior performance; with only sloughing occurring at the free face. In only two cases, model 2 tested at 0°C and model 3C where the fill temperature was increased to -0.3°C, that some structural failure was experienced. In all cases, even at elevated temperatures, no water drainage was noted, thus indicating that stopes filled with frozen limestone backfill at 5 and 10% water content should not experience water run off. Some consolidation of the fill mass, reaching 6% of the fill height, should however be expected. The simulations have indicated that limestone frozen fill, placed at 5 to 10% water content, should still provide stope structural support even when variations in temperature are experienced underground.

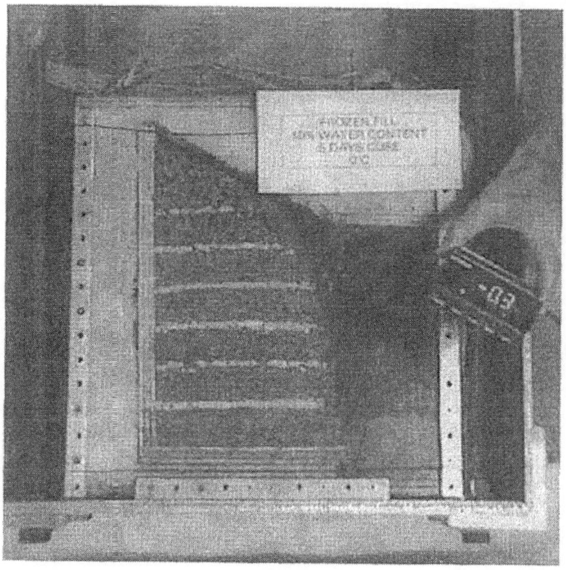

Figure 12. Limestone fill (10% water, -0.2°C), model 4B

Two model simulations were performed using the siltstone material. Model 5, prepared at 5% water content and cured at a temperature of -0.8°C, failed during loading at 8 g (Figure 13). This would correspond to a fill height of only 1.9 m. It should be however noted that failure, although reaching some 20% of the fill volume, occurred primarily close to the free face. The fill core was still frozen solid and should have continued to provide ground support at increased loading. Model 6, prepared at 10% water content and cured at a temperature of -0.7°C, failed during loading at 5 g, which corresponds to a fill height of only 1.2 m. Fracturing and failure reached the core of the stope, resulting in a large volume of fill volume, some 25%, displacing toward the empty room (Figure 14).

The frozen fill modelled stopes using the siltstone material demonstrated lower performance; with structural failure occurring at the tested temperatures. Also, water drainage was noted, in particular for the model stope filled with frozen backfill at 10% water content, thus indicating the possibility of water run off in the field. The simulations have indicated that siltstone frozen fill, placed at 5 to 10% water content, may not provide stope structural support at elevated temperatures reaching -0.7°C.

Figure 13. Siltstone fill (5% w.c., -0.8°C), model 5

Figure 14. Siltstone fill (10% w.c., -0.7°C), model 6

6 CONCLUSIONS

The static strength test work and dynamic model simulations presented in this paper indicates that the frozen fill materials exhibit superior structural performance, in particular at lower temperatures. The fill recipes were prepared at lower than typical water contents (5 and 10% by weight water content) to evaluate the possibility of minimizing water run off experienced in the field. The limestone frozen fill demonstrated the capability to provide stope structural support even when variations in temperature are experienced underground (from -12.9°C to -0.2°C). The siltstone frozen fill, however, did not provide the required stope structural support at elevated temperatures reaching -0.7°C. It is noted that, while both fill materials exhibited comparable uniaxial compressive strengths, their dynamic performance was much dissimilar.

REFERENCES

Andersland, O.B. & Anderson, D.M. 1978. *Geotechnical Engineering for Cold Regions*. McGraw-Hill.

Archibald, J.F. & Nantel, J.H. 1986. Frozen Fill for Underground Support. CIM Bulletin. 79(885):45-49.

Goughnour, R.R. & Andersland, O.B. 1968. Mechanical Properties of a Sand-Ice System. *Journal Soil Mech. Found. Division of American Society of Civil Engineers.*

Kelly, M.D. 1983. *Investigation of the Strength Characteristics of Frozen Backfill for Use in Underground Mines Situated in Permafrost Zones*. M.Sc. Thesis. Department of Mining Engineering, Queen's University.

Mitchell, R.J. 1991. Centrifuge modelling as a consulting tool. *Canadian Geotechnical Journal*. 28(1): 162-167.

Sprott, D. 1983. *Ice as Backfill Material in Underground Mines*. M.Sc. Thesis. Department of Mining Engineering, Queen's University.

Tsytovick, N.A. 1975. *The Mechanics of Frozen Ground*. McGraw-Hill.

Mining in the Arctic, Olsen, Lorentzen & Rendal (eds), © 2001 Swets & Zeitlinger, Lisse, ISBN 90 5809 177 5

Ground control monitoring at the Nanisivik Mine

N. L. Lecuyer
Manager of Technical Services, Breakwater Resources Ltd., Nanisivik Mine

ABSTRACT: The Nanisivik Mine is located on the 73rd parallel in the Canadian Arctic. The mine began commercial production in 1976 and has since produced over 16 million tonnes. The Main Ore Zone stretches for 3.5 km west to east and has an average width of 80 m with a height of 5-20 m. The room and pillar mining method was employed in the Main Ore Zone with approximately 800,000 tonnes remaining in pillars. During 1999, post pillar recovery commenced as an integral part of the annual mine production. Preparation for post pillar recovery was initiated a decade ago with a three phase study that addressed ground stress determinations, ground stability evaluation and a pillar recovery trial. The major pillar recovery program commenced after Breakwater Resources acquired the mine in 1996. Tools such as the PIN (pillar index number) system and GMM (ground movement monitors) were developed and implemented to assist in the ground control aspect of post pillar recovery. This paper reviews the post pillar recovery plans as well as the monitoring systems utilized to ensure maximum pillar recovery coupled with maximum safety for the employees.

1 INTRODUCTION

The Nanisivik Mine is located 750 km north of the Arctic Circle at the north end of Baffin Island.

Located on Strathcona Sound, the mine has access to the Arctic Ocean by way of Admiralty Inlet and Lancaster Sound. A deep sea port allows for unloading of cargo and loading of concentrate stored at the dock area. (Figure 1)

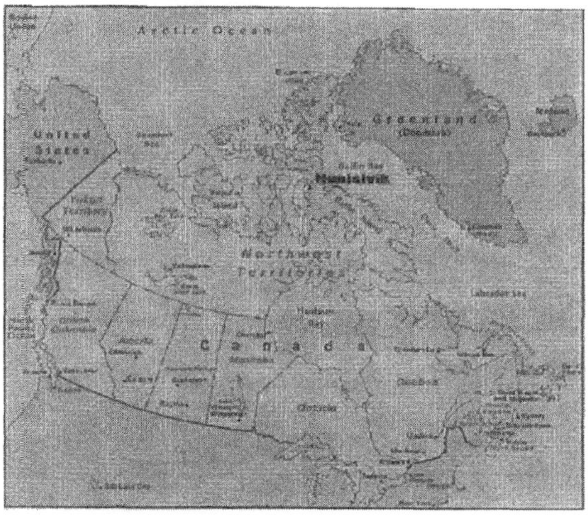

Figure 1-Location Map

The Nanisivik Mine has been in production since 1976 with over 16 million tonnes of ore mined to date. The present ore reserves indicate a mine life of approximately four years. The installation of a dense media separation plant, scheduled for start up in mid 2001, will permit mining and upgrading of diluted ore material offering an extension to the mine life.

The mine production is currently 2,500 t/day of zinc-lead ore with several mining methods utilized. The Main Ore Zone or MOZ was mined by the room and pillar method and is composed principally of pillars of good grade whose mixing is an important part of the annual mine production.

2. GEOLOGY

The mineralized deposits at Nanisivik are carbonate hosted, similar to Mississippi Valley type deposits in a block faulted sediment sequence of Proterozoic age. The main geological units are the dolostone of the Upper Society Cliffs formation which hosts the mineralization and the shale and dolostone of the overlying Victor Bay formation. (Figure 2)

Extensive faulting has resulted in a horst and graben structural setting. The mineralization formed by the replacement of the host rock in two main sulphide events, an early barren massive pyrite-dolomite event (uneconomic) and a later event, pyrite-

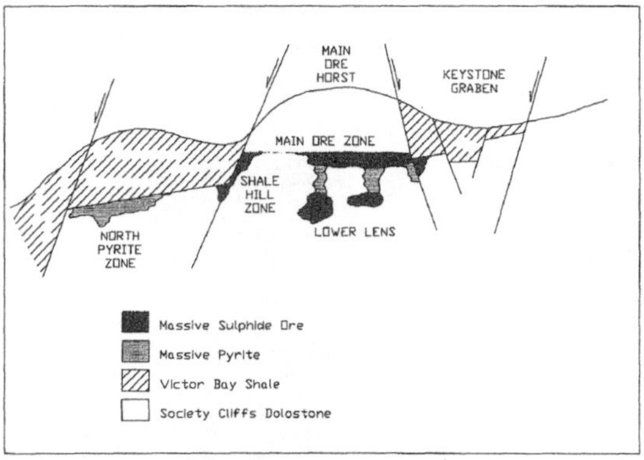

Figure 2 –Schematic Geologic Section looking east.

sphalerite-minor galena-dolomite, that produced the economic mineralization. The Nanisivik deposit is a major sulphide deposit with an estimated 50-100 million tonnes of massive pyrite of which about 15 million tonnes being economic due to zinc enrichment.

3. MINING OPERATIONS

In the Nanisivik area, the rock is permanently frozen to depths of 500 m or more, which contributes to mine stability. The MOZ is on average 80 m in width, over 3,500 m in length by 5-20 m in height.

This primary mining method is room and pillar.

The room and pillar design includes pillars of 5 by 5 to 7 by 7 m on 25 m centers. The Main Ore Zone has been largely mined out with approximately 80% recovery on the first pass. The MOZ outcrops at the west and east ends of the orebody allowing access and egress via portals. The MOZ provided most of the production in the past, and consists mostly of post pillars today with about 75% of the MOZ ore reserves contained in the pillars.

Production comes mainly from the satellite zones located to the north of and beneath the Main Lens. Figure 3 illustrates the extent of the mine.

The mining methods utilized in the satellite zones include longhole , benching, slashing, and drifting, all of which make up 70% of the total mine production. Limited open pit mining is also carried out but this only amounts to about 5% of the total mineable ore reserves. All development and production drilling is done dry due to the permafrost. The rock temperature is a constant –12°C and the silica content is low, enabling maintaining respirable dust limits using dust collectors on the drilling equipment. Blasting is done using ANFO and non-electric detonators. ANFO is prepared on site with loading done using a mobile loading unit.

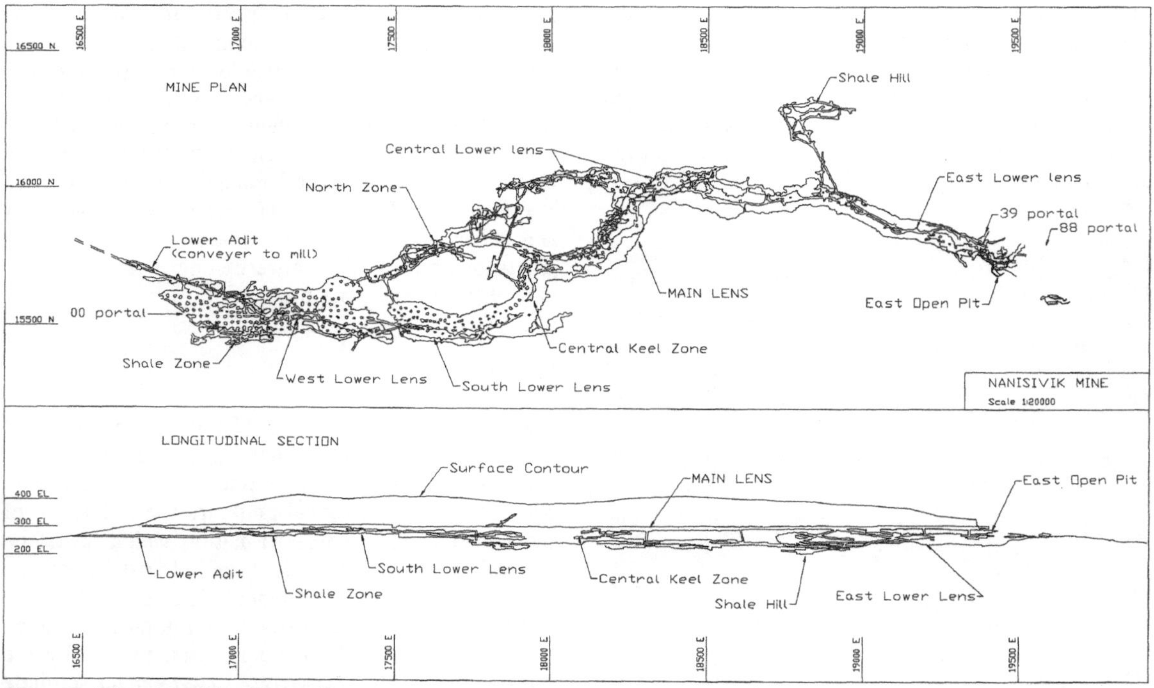

Figure 3- Mine plan & longitudinal section.

4. PILLAR RECOVERY

The shallow depth of the orebody coupled with the permafrost present excellent conditions for post pillar recovery. Preparation for pillar recovery was started in 1989 with studies by CANMET. Pillar recovery operations ensued to some degree until 1994 when a large ground fall brought the program to a halt. Following this, a review of the pillar recovery program was carried out in 1997 and a revised program was developed that would provide a safe and economical method for pillar extraction. Presently some 24% of the total ore reserves are contained in the MOZ pillars, making successful recovery of the pillars essential for the viability of the mine.

5. ROCK MECHANICS STUDIES

In 1989, CANMET undertook rock mechanic studies with the objective to achieve pillar recovery without serious deterioration occurring in the roof (back) and wall strata. CANMET was retained to carry out a three- phase program between 1989 and 1991. Phase one, included field investigations and laboratory testing to determine ground stresses, insitu deformation modulii and borehole inspections. Measurements at the East Portal of the mine determined pre-mining vertical stress and stresses perpendicular and parallel to the horizontal orebody. These stresses can be expressed as follows:

Vertical Stress : 0.028 Mpa/m (1)
Stresses $\perp$ to Orebody : 3.2 x vertical stress. (2)
Stresses $=$ to Orebody : 1.4 x vertical stress (3)

Also, deformation modulii in the range of 50-60 Gpa for the dolomite and 20-30 Gpa for the orebody were recommended. Video inspection up to 6 m into the roof (back) indicated sound roof conditions aside from blast damage close to the roof line. From the modulii determined (elastic modulus and deformation modulii), the rock mass rating or RMR was evaluated at 79 for the dolostone and 60 for the sulphide ore.
Phase two consisted of ground stability evaluations utilizing finite element simulations to estimate overall mine stability. The parameters examined included crown pillar stability, sill pillar stability, pillar mining sequence and stope stability. These simulations showed that:

1.) The complete removal of the post pillars would be feasible without the use of backfill.
2.) The crown pillar was stable.
3.) The sill pillars were stable, but could require some ground control measures.
4.) From the point of view of access and ground control, pillar recovery from east to west was best.

Phase three included a pillar recovery trial carried out at the East Portal area of the mine. (Figure 4) In all, five pillars were extracted and observations and field instrumentation confirmed ground stability consistent with earlier analytical assessments. Back analysis and monitoring data were used to validate the model and input parameters used. Also, field abutment and pillar stresses determined using strain rings were compared with those predicted by the numerical model simulations and the results were in good agreement. The results supported the feasibility of extracting all of the remaining post pillars.
Another 11 pillars were removed in this location leaving an unsupported area of 350 m by 50 m.
Following the rock mechanics studies, instrumentation was installed between blocks 18 and 26. Between late 1992 and 1994 some 21 pillars were mined. In February of 1994 after six pillars had been removed in the 20 block area there was a large ground fall (approximately 12,000T) that halted pillar mining completely.

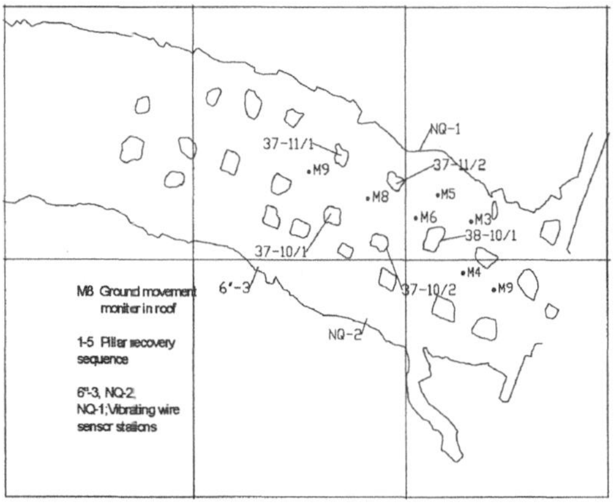

Figure 4 – Test site stope & monitoring layout.

6. PILLAR RECOVERY PLAN

Following the 1994 ground fall in the 20 block a review of the mine design was required in order to satisfy the Mines Inspection Services before pillar recovery could continue. During 1995 and 1996 no pillar mining was done and meeting grade objectives became difficult without the pillar tonnage.

In early 1997, the Nanisivik staff undertook a review of the mine design and the pillar recovery procedures under the direction of Dr. Chris Page, principle of the firm Steffen, Robertson and Kirsten(Canada) Inc.
Over the 24 years of mine operation, eight major ground falls have been documented and all, with the exception of the 20 block ground fall in 1994, have occurred in areas in which man entry was possible. Also, all of the ground falls had a visual structural explanation including exposure of a wedge or a shallow angle slip. There have been no ground falls due to failure of an intact beam. The conditions that had caused the ground falls were well understood. The pillar extraction program was therefore addressed with the critical strategy being that of two types of entry. These were defined as ENTRY: where conditions of geometry and rock mass quality allowed for the safe working of men within the current excavation and NON-ENTRY: where the conditions of geometry and rock mass were not sufficiently reliable for man-entry.
The primary concern for pillar extraction was safety, ensuring that men were only allowed in areas where the stability was reasonably certain. The major concerns included roof falls in Entry areas, collapses in Non-Entry areas that could overrun the last line of pillars, collapses in the Non-Entry areas that could trigger air blasts and caving in Non-Entry areas that could cause major collapses in Entry areas. The layout criteria for the room and pillar stoping was 5 x 5 to 7 x 7 m pillars on 25 m centers for the dolostone or mineralized backs and 5 m thick rib pillars on 20 m centers for the backs in the shale zone areas.
The evaluation of excavation stability was based on the use of the rock mass rating (RMR) as this was considered to be a good empirical procedure for assessing the excavation stability. (Figure 5)

Using the Hydraulic Radius, the RMR values could be calculated for Entry type excavations and the RMR at the possible start of caving. While the empirical relationship is approximate at best, with experience, it was considered a reliable tool. Therefore to provide limits to work with until more information could be acquired, the following criteria could be used to distinguish Entry from Non-Entry mining and also to limit the geometry of the pillar extraction areas.

1.) A Hydraulic Radius of approximately 30 could be developed with a competent dolostone back without large ground falls occurring.
2.) A Hydraulic Radius of approximately 10 could result in large failures with the shales, and
3.) Caving might occur at Hydraulic Radius of over 40.

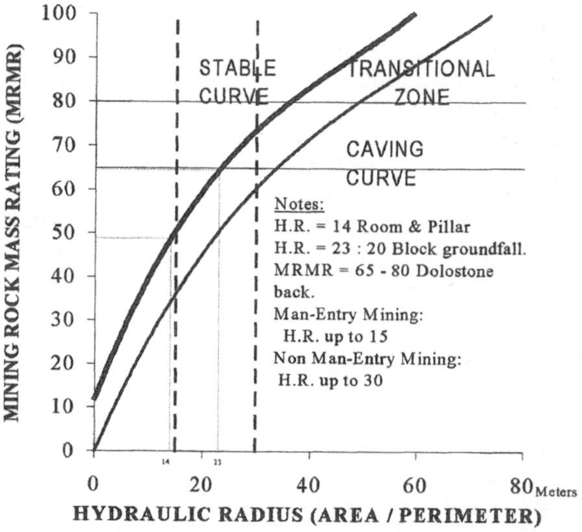

Figure 5- Stability Diagram (after Dr. D. H. Laubscher)

From this, the following controls were established. Entry mining, with average dolostone backs,could be defined by a Hydraulic Radius of less than 15. Entry mining in shale could be defined by a Hydraulic Radius of 5, as was presently being done.
The definition of an Entry Area was that within the limits of the present room and pillar design; only remote equipment beyond the last pillar line. Exemption from this would be approved on each individual occasion and only if the Hydraulic Radius and the RMR values would plot within the stable zone. Non-Entry mining with average dolostone backs was to be limited by a Hydraulic Radius of 30, while for shale backs this would be limited to 10.

7 PIN CLASSIFICATION

A pillar index number, or PIN system, was developed at Nanisivik during 1997 to compare different pillars and provide a measure for deciding the sequence in which pillars should be extracted and /or which might be left for a longer period or permanently. The PIN evaluation was based on the following factors with each factor having points assigned.

1.) Grade-%Zn -higher grade more points :max. 1 point.

2.) H/W Ratio-higher the height/width ratio higher points to max. of 2.
3.) Profile- above average to poor, with poor assigned higher points to max. of 1.5
4.) Jointing- massive, minor, major with major jointing assigned most points to max. of 1.5
5.) Weathering- none, minor, major with major assigned most points to max. of 1
6.) Structure- none, minor, major with major assigned most points to max. of 3

All of the 385 pillars in the Main Lens were evaluated and rated. Therefore, economic and geotechnical factors were used to rate the pillars. In general, a pillar with a high rating would be more attractive to recover before a similar pillar with a lower rating. The more competent pillars have a lower PIN and thus should be left in place as long as possible. In general terms, the average PIN was 2.5 with the interpretation that pillars with a PIN of less than 2 were not very attractive to recover and those with a PIN of over 3 were. From this information, plans could then be color-coded to aid in visualizing the pillar excavation priorities.

8 MONITORING

During the mine design review, it was decided to invest in real-time monitoring of areas to provide an additional tool to aid the rock mechanics program.
For this, ground movement monitors or GMM's (extensometers) were selected and CANMET was contracted to help Nanisivik in setting up the system. A system was installed, capable of having a total of 32 GMM units in the roof (back) to monitor any downward block movement. The GMM units consisted of Baytech Model ETT-204 electronic telltales. Four stations or slaves were purchased and installed underground with each slave capable of having 8 units operating from it. The GMM's consisted of 2.44m rockbolts (16mm diameter) coupled together to give a total length of 10 m. A standard 50 mm shell anchor was used at the end of the hole and the ETT-204 installed at the collar with a standard collar plate. (Figure 6)
A surface master and computer were set in the mine office. Software called LOOKOUT was installed which accesses the slaves via the master and monitors, displays, logs and provides alarms based on preset trigger levels. Movement beyond the trigger levels initiates the alarm, which can be verified. A log book is kept in the mine office which is used to record any changes that require follow-up by the electrical department. Also a ground control logbook is utilized for entry of any movement so that the proper protocol is followed. Alarm procedures are established so that any situation may be investigated immediately and corrective action taken.

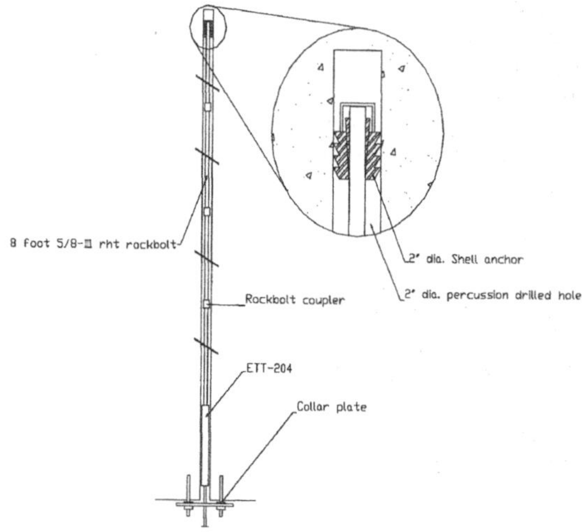

Figure 6- GMM Installation

The underground leaky feeder communication system serves to transmit the data through to the computer on surface. All stations are indicated on the computer monitor and readings can be verified over the last hour, 12hour and weekly periods. Also a plot of the displacement versus time can be viewed for any GMM via a push button.

9 RESULTS TO DATE

Since the review of the mine design and pillar extraction procedure in 1997, pillar recovery has progressed on a continuous basis. Pillar recovery has been very good, with 96% of the forecasted tonnage recovered in 1998, 90% during the 1999 period and to date 2000 about 75%. The 2000 program is on schedule because recovery of ore in the north wall (gouging) is also being done as final retreat operations are in progress.
While the GMM units are by no means trouble free, they provide continued monitoring. Installed initially in Entry areas, some units are now located in Non-Entry areas and provide ongoing monitoring of maximum span openings. Open spans have attained 80 m in the MOZ-Central area and 70 m in the MOZ-East area of the mine. Careful blasting has helped in maintaining the GMM units in open areas. The durability of these units over time remains to be seen. The data transmitted from the units is dependant on the mine radio (leaky feeder system) and electrical noise can be a problem.

Since 1997, several ground falls have occurred. In the MOZ-Central, a ground fall occurred in September of 1999 of about 5,000 tonnes. Another ground fall occurred in October of 2000, which was estimated at 10-12,000 tonnes. (Figure 7)

Both of these ground falls were a result of sulphides left in the roof (back) where it was known that low angle slips occurred. This particular area of the mine was also exposed to warmer air in the past as the main ventilation system downcasted the 17N ramp. The compressor station was installed in this area at the time resulting in a source of warm air. Geological features included the presence of vuggy ground as well as low angled slips. This area had not been mined to the dolostone contact. Both of the ground falls were contained in the Non-Entry areas hence there was no danger to mine personnel. The ground falls occurred shortly after pillar recovery in the immediate area. While not accurately predictable, the ground falls were anticipated due to the conditions mentioned.

While the GMM's did not indicate any movement prior to the September 1999 ground fall, the GMM B-4 did indicate 0.40 mm of movement the day prior to the October 2000 ground fall. In other areas such as 19 block where five pillars were recently removed the GMM did show slight movement but stabilized very quickly. Presently, eight GMM's are located in Non-Entry areas and eleven are located in Entry areas. These are located along travelways and therefore continue to provide information on the ground movement in these areas. The GMM's, while providing useful information, have to be utilized with caution since each only represents a measurement at a single point.

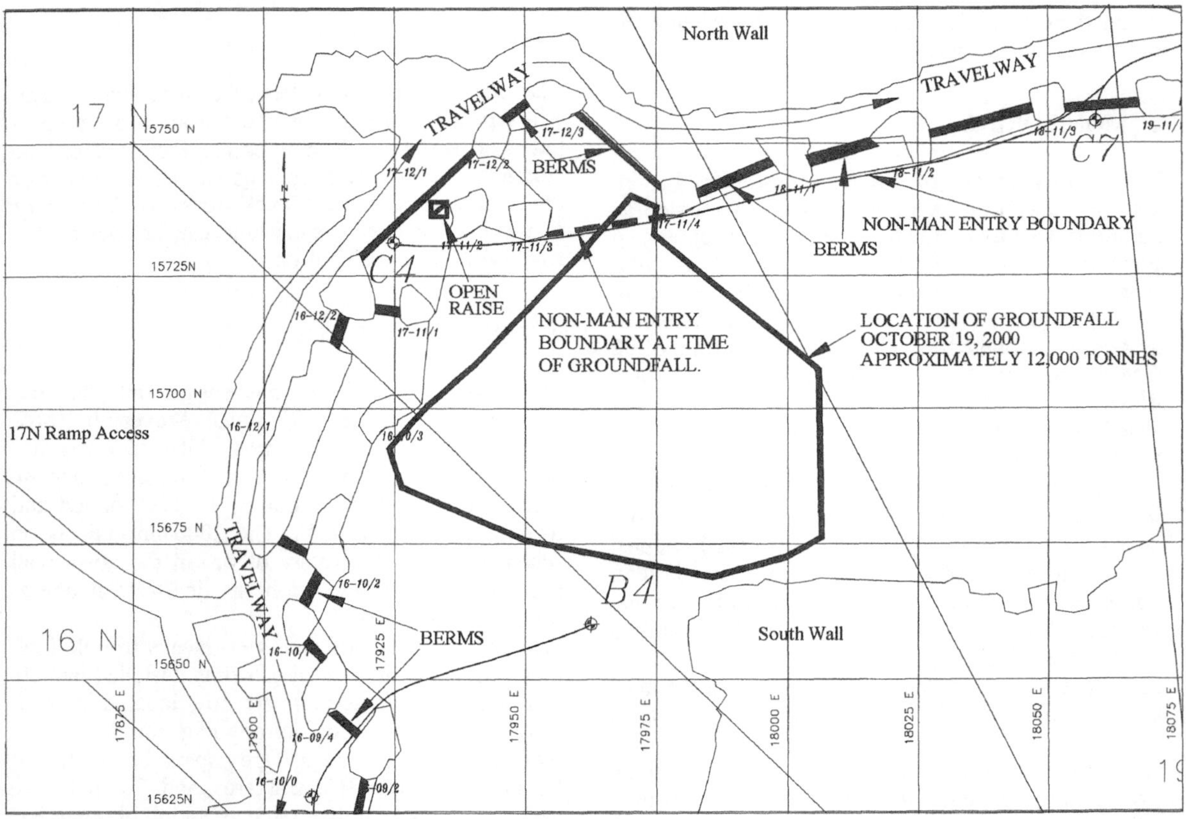

Figure 7- 17 Block Area Ground Fall

10 ADDITIONAL MONITORING

For future pillar recovery in the MOZ-West part of the mine additional monitoring is being considered due to the presence of the Keystone faults. The MOZ-West has approximately 300,000 tonnes of ore in pillars grading over 11% zinc. (Figure 8)
Removal of pillars would mean removing support with potential downward movement. In order to monitor any movement, a program is being investigated with CANMET to utilize Time Domain Reflectometry or TDR. This would expand the surface based mine stability monitoring program with the addition of TDR cables to the existing GMM array. This would provide real-time monitoring of fault movement during and after pillar recovery, help decision making with respect to man entry and worker safety in pillar recovery areas and improve the pillar recovery program by enhancing the decision making processes. The TDR program is expected to require about six months to set up.

11 CONCLUSIONS

The experience to date has provided information that permits formulation of the following conclusions.

1.) The design guidelines for Hydraulic Radius have reached 30 with no indication of competent and regular roofs (backs) showing signs of beam failure. Therefore there is no need at the present time to change the guidel lines.

2.) No significant signs of stress related failure from pillar walls have been observed. The ground falls that have occurred are a result of exposure of a wedge of shallow angle slip. Any sulphides left in the roof (back) can be prone to a ground fall as the sulphides break away from the dolostone contact if the span becomes greater than the normal room and pillar spans.

3.) The GMM monitoring system provides real-time indication of ground movement. While it has not served to directly predict the ground falls that have occurred, the extensometers provide assurance that where installed, conditions in the roof (back) are monitored. These are point load type extensometers that serve more to monitor conditions over the longer term with respect to progressive deterioration of the roof (back) conditions. The system provides the operator with physical indications of movement.

4.) The use of Entry and Non-Entry pillar recovery procedures has proved to be successful at Nanisivik and will continue to be utilized, providing for maximum safety and post pillar recovery.

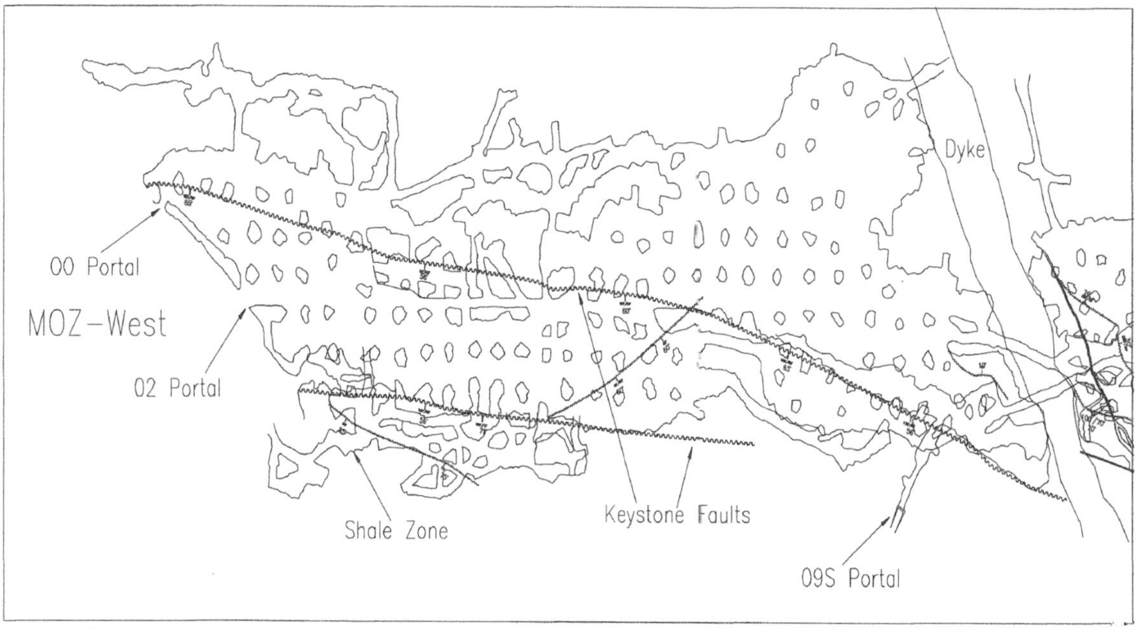

Figure 8- MOZ-West area.

REFERENCES

Arjang,B. & Herget,G. 1989. Post Pillar Recovery Strategies at Nanisivik Mine, Phase 1: Ground Control Instrumentation & Stress Determinations. MRL89-147 (TR)

Judge, K. 1998. Ground Movement Monitoring System, CanZinco's Nanisivik Mine, Mining and Mineral Science Laboratories, Canmet, MMSL 98-038 (CR), Job no. 601689

Page, C. 1999. Review Visit, Pillar Extraction Program, 24-26 July,1999, Steffen, Robertson & Kirsten (Canada) Inc.

Page, C. 1997. Site Visit Jan. 17-23. Steffen, Robertson & Kirsten (Canada) Inc.

Vongpaisal, S. & Udd, J.E. & Larocque, G.E. 1990. Post Pillar Recovery Strategies at Nanisivik Mine, Phase 2: Ground Stability Evaluation MRL 90-27 (TR)

Vongpaisal, S. & Herget, G. & Udd, J.E. & Larocque, G.E. & Bailey, F. 1991. Post Pillar Recovery Strategies at Nanisivik Mine, Phase 3: Pillar Recovery Trial MRL 91-123 (TR)

Tremblay, S. 1998. Aspects Economique et Geotechnique du programme de recuperation des piliers par l'application du Pillar Index Number. Project de fin d'etude.

Mining in the Arctic, Olsen, Lorentzen & Rendal (eds), © 2001 Swets & Zeitlinger, Lisse, ISBN 90 5809 177 5

Author index